Lecture Notes in Physics

330

J. M. Moran J. N. Hewitt K. Y. Lo (Eds.)

Gravitational Lenses

Proceedings of a Conference Held at the
Massachusetts Institute of Technology
Cambridge, Massachusetts, in Honour of
Bernard F. Burke's 60th Birthday, June 20, 1988

Springer-Verlag
Berlin Heidelberg GmbH

Editors

J. M. Moran
Harvard-Smithsonian Center for Astrophysics
60 Garden Street, Cambridge, MA 02138, USA

J. N. Hewitt
Haystack Observatory
Off Route 40, Westford, MA 01886, USA

K. Y. Lo
Department of Astronomy, University of Illinois
1011 W. Springfield Avenue, Urbana, IL 61801, USA

ISBN 978-3-662-13731-4 ISBN 978-3-540-46142-5 (eBook)
DOI 10.1007/978-3-540-46142-5

Originally published by Springer-Verlag Berlin Heidelberg New York in 1989
Softcover reprint of the hardcover 1st edition 1989

2158/3140-543210 – Printed on acid-free paper

Bernard Burke with some of his students and former students

Front row (L to R): Alan Rogers, Martin Ewing, (Bernard Burke), Jacqueline Hewitt, Thomas Wilson, Fred Lo. Back: Hans Hinteregger, Keith Tuson, Michael Heflin, Antonio Garcia-Barreto, Charles Lawrence, Samuel Conner, Joseph Lehár, James Moran, Glen Langston, Vivek Dhawan

Organizing Committee

C. C. Barrett, CfA
G. W. Clark, MIT
J. N. Hewitt, Haystack
K. Y. Lo, U. Illinois
J. M. Moran, CfA (Chair)
A. E. E. Rogers, Haystack
V. Romansky, MIT
I. I. Shapiro, CfA
E. L. Turner, Princeton
T. L. Wilson, MPIfR

Preface

This volume contains the papers presented at a Conference on Gravitational Lenses held at the Massachusetts Institute of Technology on June 20, 1988, in honor of Bernard Burke's 60th birthday. The idea for this meeting arose in a late-night discussion among Fred Lo, Jim Moran, and Tom Wilson. The topic of gravitational lenses quickly became the obvious choice since Bernie was one of the pioneers of the field and has devoted much of his creative energy to it over the past decade. As Roger Blandford said in his talk, "It is characteristic of [Bernie] that the subject that we are discussing here is not some half-forgotten, thirty-year-old scientific triumph but instead a topic that has become even more exciting since this meeting was announced."

About 80 people attended the conference, and 27 papers were presented, all of which are in this volume. About half of Bernie's nineteen Ph.D. students were in attendance. George Clark began the conference with a sketch of Bernie's scientific career. Dennis Walsh then described the little-known events that led to the discovery of the first gravitational lens 0957+561.

The scientific sessions were followed by a reception and banquet held at the Hyatt Regency Hotel, which was attended by Bernie's family, conference participants, and many friends and colleagues. At least four generations of students were present, including Bernie's supervisees, his supervisor, and his supervisor's supervisor. After the banquet, tributes were paid by Al Hill, former Vice President of Research at MIT and one of Bernie's earliest mentors; Woody Strandberg, Bernie's thesis supervisor; Irwin Shapiro, Director of the Harvard–Smithsonian Center for Astrophysics; Robert Hughes, President of Associated Universities; Paul Vanden Bout, Director of the National Radio Astronomy Observatory; Herman Feshbach, former Chairman of the Physics Department at MIT; Paul Sebring, former Director of Haystack Observatory; and Martin Ewing and Charles Lawrence, two of Bernie's Ph.D. students. Several gifts were presented, which should keep Bernie from getting lost in the fog around Marblehead harbor.

This conference was sponsored by the MIT Physics Department and Haystack Observatory with additional support from the Research Laboratory of Electronics and the Harvard–Smithsonian Center for Astrophysics. We are grateful to Carolann Barrett, Kathleen McCue, Alan Rogers, and Veronica Romansky for their help with the local arrangements. John Cook of RLE took the photographs shown here. Roger Blandford offered much helpful advice on the proceedings. Mark Birkinshaw provided important editorial assistance, including the difficult task of transcribing the tape recording of Dennis Walsh's talk. The final form of these proceedings is due in large measure to the careful work of Carolann Barrett, who served as typist and copyeditor.

J. M. Moran
J. N. Hewitt
K. Y. Lo

December 28, 1988

Table of Contents

Page

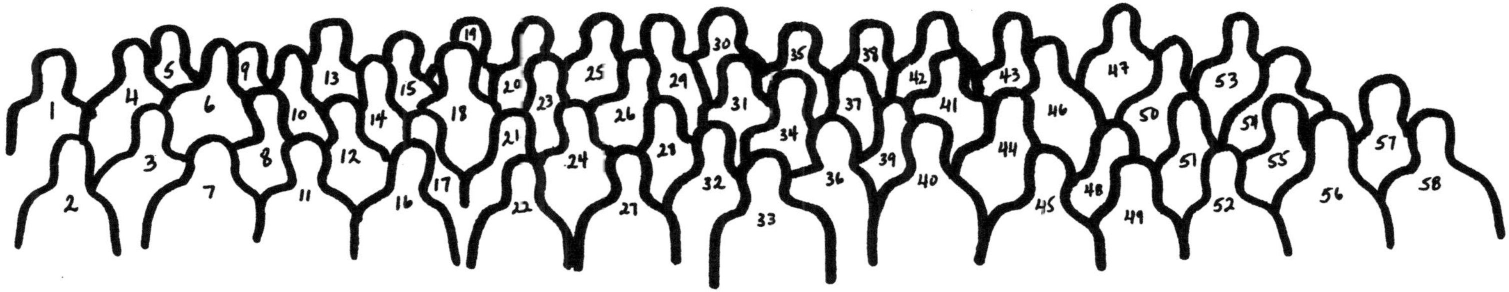

1. Vivek Dhawan 2. Michael Heflin 3. Irwin Shapiro 4. Keith Tuson 5. Samuel Conner 6. Jason Bochinsky 7. Luis Rodríguez 8. Alan Rogers 9. Genevieve Soucail 10. Glen Langston 11. Joseph Lehár 12. J. Antonio Garcia-Barreto 13. Joseph Binsack 14. Marc Gorenstein 15. Michael Ratner 16. Roger Cappallo 17. Mark Reid 18. Norbert Bartel 19. Brian Corey 20. David Smith 21. Claude Canizares 22. Peng Yun-Lou 23. Russell Lavery 24. William Watson 25. Hans Hinteregger 26. Roger Blandford 27. Li Gang 28. Martin Ewing 29. Mark Birkinshaw 30. Karl Menten 31. Edwin Turner 32. Rachel Webster 33. Bernard Burke 34. Charles Lawrence 35. Bernard Feld 36. Jacqueline Hewitt 37. Leonid Ozernoy 38. Peter Schneider 39. Dennis Walsh 40. James Moran 41. Thomas Wilson 42. Vahé Petrosian 43. Joachim Wambsganss 44. George Clark 45. Fred Lo 46. Richard Porcas 47. Alan Whitney 48. Irving Segal 49. Joseph Salah 50. Colin Lonsdale 51. Robert Hughes 52. Paul Vanden Bout 53. Gary Hinshaw 54. John Dreher 55. James Wright 56. Jeffrey Isaacson 57. Scott Grossman 58. Hans-Walter Rix

List of Participants

Alan H. Barrett, Massachusetts Institute of Technology, Cambridge, MA
John W. Barrett, Massachusetts Institute of Technology, Cambridge, MA
Norbert H. Bartel, Center for Astrophysics, Cambridge, MA
Joseph H. Binsack, Massachusetts Institute of Technology, Cambridge, MA
Mark Birkinshaw, Center for Astrophysics, Cambridge, MA
Roger D. Blandford, California Institute of Technology, Pasadena, CA
Hale V. Bradt, Massachusetts Institute of Technology, Cambridge, MA
James Brody, Haystack Observatory, Westford, MA
Bernard F. Burke, Massachusetts Institute of Technology, Cambridge, MA
Claude R. Canizares, Massachusetts Institute of Technology, Cambridge, MA
Roger Cappallo, Haystack Observatory, Westford, MA
George W. Clark, Massachusetts Institute of Technology, Cambridge, MA
Samuel Conner, Massachusetts Institute of Technology, Cambridge, MA
Susan Cooper, Massachusetts Institute of Technology, Cambridge, MA
Brian Corey, Haystack Observatory, Westford, MA
Peter T. Demos, Massachusetts Institute of Technology, Cambridge, MA
Vivek Dhawan, Center for Astrophysics, Cambridge, MA
John W. Dreher, Massachusetts Institute of Technology, Cambridge, MA
Martin S. Ewing, California Institute of Technology, Pasadena, CA
Bernard T. Feld, Massachusetts Institute of Technology, Cambridge, MA
Herman Feshbach, Massachusetts Institute of Technology, Cambridge, MA
David H. Frisch, Massachusetts Institute of Technology, Cambridge, MA
J. Antonio Garcia-Barreto, UNAM, Ensenada, Mexico
Marc V. Gorenstein, Millipore Corporation, Milford, MA
Scott A. Grossman, Steward Observatory, Tucson, AZ
Michael Heflin, Massachusetts Institute of Technology, Cambridge, MA
Jacqueline N. Hewitt, Haystack Observatory, Westford, MA
Albert G. Hill, Massachusetts Institute of Technology, Cambridge, MA
Gary Hinshaw, Oberlin College, Oberlin, OH
Hans F. Hinteregger, Haystack Observatory, Westford, MA
Robert Hughes, Associated Universities, Inc., Washington, DC
Jeffrey A. Isaacson, Massachusetts Institute of Technology, Cambridge, MA
Kenneth Johnson, Massachusetts Institute of Technology, Cambridge, MA
Vera Kistiakowsky, Massachusetts Institute of Technology, Cambridge, MA
Lawrence M. Krauss, Yale University, New Haven, CT
Glen Langston, Naval Research Laboratory, Washington, DC
Russell J. Lavery, Institute for Astronomy, Honolulu, HI
Charles R. Lawrence, California Institute of Technology, Pasadena, CA
Joseph Lehár, Massachusetts Institute of Technology, Cambridge, MA

Li Gang, Center for Astrophysics, Cambridge, MA
Kwok Yung Lo, University of Illinois, Urbana, IL
Colin J. Lonsdale, Haystack Observatory, Westford, MA
Francis E. Low, Massachusetts Institute of Technology, Cambridge, MA
James Mahoney, Marlboro College, Marlboro, VT
Karl M. Menten, Center for Astrophysics, Cambridge, MA
James M. Moran, Center for Astrophysics, Cambridge, MA
Philip Morrison, Massachusetts Institute of Technology, Cambridge, MA
Joel Orlen, American Academy of Arts and Sciences, Cambridge, MA
Leonid Ozernoy, Center for Astrophysics, Cambridge, MA
D. Cosmo Papa, Massachusetts Institute of Technology, Cambridge, MA
Peng Yun-Lou, Nanjing University, Nanjing, People's Republic of China
Vahé Petrosian, Stanford University, Stanford, CA
Richard W. Porcas, Max Planck Institut für Radioastronomie, Bonn, FRG
Michael I. Ratner, Center for Astrophysics, Cambridge, MA
Mark J. Reid, Center for Astrophysics, Cambridge, MA
Edward C. Reifenstein III, Compugraphics, Wilmington, MA
Hans-Walter Rix, Steward Observatory, Tucson, AZ
Luis F. Rodríguez, Center for Astrophysics, Cambridge, MA
Alan E. E. Rogers, Haystack Observatory, Westford, MA
Philip W. Rosenkranz, Massachusetts Institute of Technology, Cambridge, MA
Joseph E. Salah, Haystack Observatory, Westford, MA
Peter Schneider, Max Planck Institut für Astrophysik, Garching, Federal Republic of Germany
Paul B. Sebring, National Radio Astronomy Observatory, Charlottesville, VA
Irving E. Segal, Massachusetts Institute of Technology, Cambridge, MA
Irwin I. Shapiro, Center for Astrophysics, Cambridge, MA
David H. Smith, *Sky and Telescope*, Cambridge, MA
Genevieve Soucail, California Institute of Technology, Pasadena, CA
David H. Staelin, Massachusetts Institute of Technology, Cambridge, MA
Malcom W. P. Strandberg, Massachusetts Institute of Technology, Cambridge, MA
John L. Tonry, Massachusetts Institute of Technology, Cambridge, MA
Edwin L. Turner, Princeton University Observatory, Princeton, NJ
Keith A. Tuson, Massachusetts Institute of Technology, Cambridge, MA
Paul A. Vanden Bout, National Radio Astronomy Observatory, Charlottesville, VA
Dennis Walsh, Nuffield Radio Astronomy Laboratory, Macclesfield, Cheshire, England
Joachim Wambsganss, Max Planck Institut für Astrophysik, Garching, FRG
William D. Watson, University of Illinois, Urbana, IL
Rachel Webster, University of Toronto, Toronto, Canada
Rainer Weiss, Massachusetts Institute of Technology, Cambridge, MA
Alan R. Whitney, Haystack Observatory, Westford, MA
Thomas L. Wilson, Max Planck Institut für Radioastronomie, Bonn, FRG
James P. Wright, National Science Foundation, Washington, DC

I. Welcoming Remarks

RECOLLECTIONS OF THE CAREER OF BERNARD BURKE

George W. Clark
Massachusetts Institute of Technology
Physics Department
Cambridge, MA 02139

On behalf of the organizing committee, the MIT Physics Department, and the Division of Astrophysics, it is my pleasure and privilege to welcome you to this conference on gravitational lenses in honor of Bernard Burke, the William Burden Professor of Astrophysics, on the occasion of his sixtieth birthday.

The theme of this conference is gravitational lenses, one of the hottest astronomical topics of the present time. It reflects just one aspect of the varied works of Bernie Burke. He is first of all a pioneering scientist who has had the vision, the expertise, and the drive to be always at the cutting edge of radio astrophysics since he took his Ph.D. at MIT some 35 years ago. With a grounding in microwave spectroscopy gained in his thesis research in the laboratory of Malcom Strandberg, Bernie was prepared to implement his youthful judgment that there was scientific gold beyond the obscure foothills of radio astronomy. Moving immediately after his Ph.D. to the Carnegie Institution of Washington, he began his career in radio astronomy and soon, with Kenneth Franklin, discovered the decametric radiation from Jupiter. That made him famous before the age of 30 and brought him the Warner Prize of the American Astronomical Society.

Ten years later, after a productive decade of radio research into the structure of our own and nearby galaxies, Bernie played a key role in developing the technique of VLBI, which has been and will forever remain in one form or another a principal tool for observing the finest detail of cosmic structures. For this work, he won the Rumford Prize of the American Academy of Arts and Sciences in 1971.

The bibliography of his prolific work in the 1970's, much of it carried out with students and colleagues who are here today, contains results on water vapor and hydroxyl masers, the cosmic background, hydrogen recombination line spectroscopy, and 100 microarcsecond imaging of quasars at millimeter wavelengths.

As soon as the VLA came on line, Bernie was at the eyepiece, so to speak. In 1979, with Perry Greenfield and Dave Roberts, he went after the radio image of the double quasar, beginning one of the central themes of the past decade of his scientific work – gravitational lenses. A second theme has been the MIT–Green Bank 5-GHz survey, with which he and his students have dredged a wealth of material for detailed study with the VLA, the great optical telescopes, and the orbiting observatories of the near future. And a third theme is the scientific future – particularly the campaign to achieve VLBI in space.

This third theme exemplifies an essential feature and benefit of Bernie's professional life: the formulation and promotion of public policy for science. Ever since his discovery of radio emission from Jupiter and the Warner Prize, Bernie has been involved in the committee, media, and advisory activities that have set government and private science research policy. As a member of the National Academy of Sciences and its various committees, he has had his hand in virtually all the studies that have guided NSF and NASA funding in radio astrophysics and led to the VLA, the VLBI network, the VLBA, and the future space VLBI.

Bernie is a science politician in the best sense of that word – a scientist with a dedication to steering public science policy in the most fruitful direction and with a sure instinct of where the levers of power reside and how to pull them. Bernie has been president of the American Astronomical Society for the past two years and has jetted back and forth across the U.S. and the Atlantic Ocean at a rate that is tiring just to contemplate. He was chairman of the Astrophysics Division of the MIT Physics Department during its formative years. For many years he has worked on developing scientific working relations with Soviet scientists and seems to be almost as well known in the Soviet Academy as he is in our National Academy. Now, among other things, he is chairman of NASA's Planetary Systems Science Working Group, figuring the best strategies for detecting other planetary systems. He is also on the faculty of the International Space University.

Bernie is a person of many parts. He is a classics scholar who can read Virgil in Latin with pleasure. On occasion he can burst forth in English verse of his own composition. He is a violist and enthusiastic chamber music player. He is a yachtsman able to hold his own in the fierce competition of Marblehead races, which sometimes he wins. He is a teacher who can convey to students a sense of how one thinks creatively in science. And he is a master of the sophisticated back-of-the-envelope calculation that can set a great scientific enterprise in motion.

But even so distinguished a scientific career as that of Bernie's, with major contributions to the advance of science to his credit, may not provoke a sixtieth birthday fest like this unless he has produced a cohort of distinguished former students who look on him as their mentor and a major source of their own scientific inspiration. Bernie has had distinguished students in abundance, some 19 by recent count, and many of them occupy key positions in the radio astronomy community today. Over half of them are here to contribute to this celebration of the sixtieth anniversary of one of the most accomplished and influential astronomers of our time.

Alan Barrett Bernard Burke

Roger Blandford lecturing.

Joseph Binsack

Colin Lonsdale M. Birkinshaw J. Wambsganss C. Canizares

Li Gang Joe Binsack C. Canizares J. Wambsganss

Bernie behind a gravitational lens, courtesy of Emilio Falco.

II. Historical Notes

0957+561: THE UNPUBLISHED STORY

Dennis Walsh
University of Manchester
Nuffield Radio Astronomy Laboratories
Jodrell Bank
Macclesfield, Cheshire SK11 9DL, England

Happy birthday, Bernie, whenever it was. I also bring greetings from the Astronomer Royal to "young Bernie." I think the last time I saw you, Bernie, was at the center of the universe. Remember that? And for those who don't know the center of the universe, it's in Beijing in a park at the spot that was defined by an emperor as the middle of the middle kingdom and hence the center of the universe.

About five years ago, there was a symposium in Liège, and on the evening of the closing day, I went out with Bernie and other people for a gourmet meal, and Bernie gave me a pretty rigorous examination on how we came to find 0957+561, which was of course the first example of multiple imaging by a gravitational lens (Walsh, Carswell, and Weymann 1979); so it occurred to me that this was a good occasion to tell the story to a wider audience.

The story actually started about 1970 or 1971 when the 76-meter Mark I telescope at Jodrell Bank was upgraded and Sir Bernard Lovell asked the members of the staff to propose new programs. I proposed to do a survey to measure accurate radio source positions for identification purposes. A student of mine, Ian Browne, had been doing this kind of work for a year or two using the interferometer at Defford, which was built by J. S. Hey and his colleagues, and which gave 1-arcsecond positions on *unresolved* sources, which permitted unambiguous identifications. This was the first time that really unambiguous identifications had been possible. In 1970, source identifications were still a very important objective of radio astronomy. There were few identified sources, and there were only a very few incomplete samples; 3C had only partly been worked through. That was the time before the Cambridge 5-km and Westerbork, and long before the VLA came into operation. Routine measurements of a second of arc were not common. They practically didn't even exist. The work done with the Defford interferometer showed that you could do this, but by then it was running out of things to do. There were two 25-meter dishes and they had observed almost everything possible within their capabilities. At Jodrell, we had the Mark IA telescope of 76-meter diameter, with the smaller Mark II, of

25-meter diameter; together these would make an interferometer with better sensitivity and better primary resolving power. I thought these could be used to find accurate positions.

We actually started in November 1972. I was working with Ted Daintree and Ian Browne. In addition, I had a new student, Richard Porcas, who lived with this project for a couple of months and did the donkey work for us. The idea was that with the Mark IA at 966 MHz we would do a finding survey consisting of raster scans for declinations 40° to 70°. As I go along, I'll point along several things that were improbable, lucky, whatever you want; you might think at the end of it that we really had no right to find the object 0957+561. But maybe that's the way science works. We had an allocation of about a month on the Mark IA telescope, and there was very high pressure, very high demand for the telescope. It was estimated that we could do the raster scan survey in three weeks. It took twice as long as we expected because we had interference from satellites and other things. At the end of our allocation, which came at the end of December 1972, we had reached declination 55°. However, Sir Bernard was very interested in what we were doing, looked at it, liked what he saw, and said, "Right, you can have another month." We carried on for another three weeks into 1973. Without this extension, our survey would have stopped just short of declination 56.°1. Lucky break number one. (As a footnote, another professor at Jodrell Bank, who was anxious to use the Mark IA to pursue his investigations into pulsars, was quite annoyed about this delay in his plans. Fortunately, he doesn't bear a grudge, and we remain friends; he is now Astronomer Royal and director of Jodrell Bank, and I owe my attendance at this meeting, in part, to him.)

The first observation that started me down this long trail was the survey scans. Figure 1 shows the first detection of the object on several adjacent declination scans. Now this was not a particularly exciting event; 0957+561 was just one of many sources, and not a very distinguished one for any obvious reason. This was on 4 January 1973, just after we got our extension. So that was the first observation, six years before it came to public notice as a gravitationally lensed object. The rate at which sources in the survey increased led Richard and Ted to devise an automatic, routine searching method. This produced a finding list of about 800 sources with typical positional accuracy of about 2 arcminutes.

We went through them all, with the Mark IA–Mark II interferometer to measure the accurate positions. We started that in June 1973. Another student, Anne Treverton (soon to become Anne Cohen), started to work on the program soon afterwards. We made the first identifications around the end of 1973. Richard did them on the Palomar Sky Survey before we

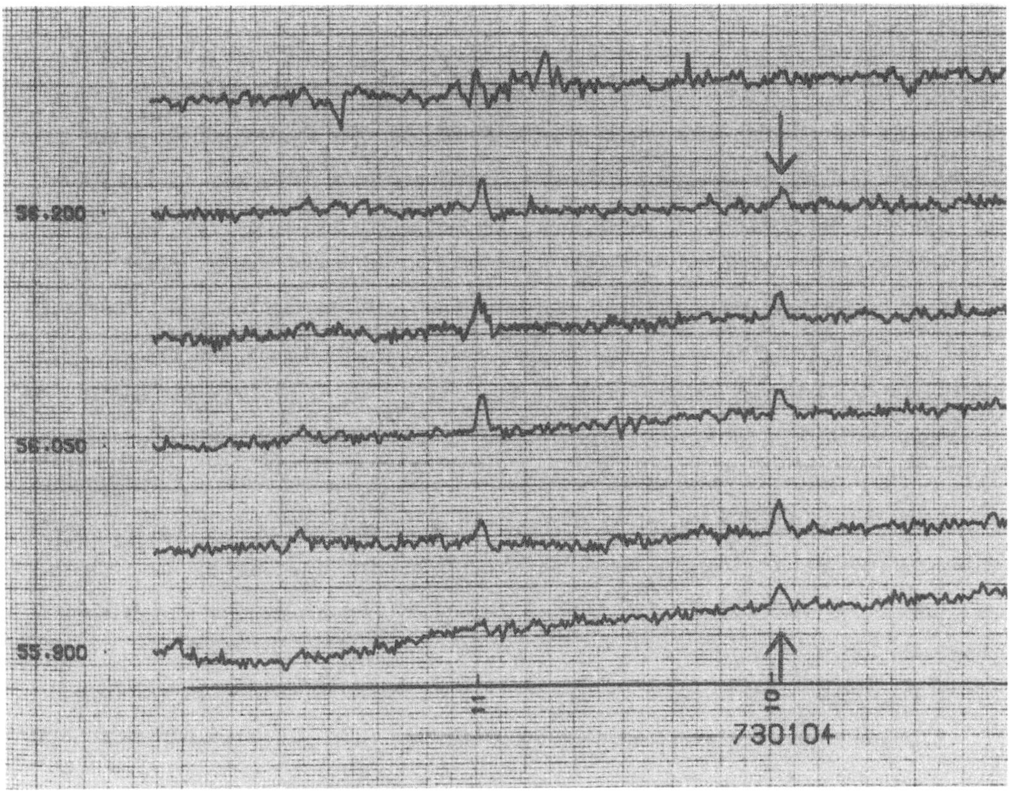

Fig. 1. Survey scans taken at the Mark IA telescope, right ascension range 09^h–12^h, declinations indicated, 0958+56 indicated by arrows.

had an accurate way of measuring the survey prints. Anne carried on later. Richard submitted his thesis in February 1975 and went off to NRAO. Anne Cohen carried out the remaining identifications, by which time we had an accurate measuring engine and could measure optical positions to half an arcsecond (Cohen *et al.* 1977). We got a final radio position accuracy of better than 2 arcseconds for about 70 percent of the sources, and we could do unambiguous identifications. The reason we could only do 70 percent instead of all of them was because some sources were resolved on that baseline or were confused by other sources within the primary beam. So it left about 30 percent for which we couldn't measure positions very accurately and so we couldn't get unambiguous identifications. The source 0958+56 (as it was originally listed in Richard's thesis) was one of these.

The program that Richard went to NRAO to do was to measure the positions of these sources using the 300-foot dish at wavelengths 6 and 11 cm. The primary beamwidth at six centimeter wavelength was 1.8 arcmin, and this was good enough to get positions to 5–10 arcseconds. That isn't really good enough for unambiguous identifications, but it's pretty useful. I want to show you some results that came essentially from Richard's work. Figure 2(a) is reproduced from the Palomar Sky Survey, and the identification of 0957+561 is indicated. It doesn't look like a double object on the scale of Figure 2(a). The dominant object in the field shown is an 11th magnitude galaxy, NGC3079. The separation of 0957+561 and NGC3079 is 14 arcminutes, and our Mark IA survey with an 18 arcmin beam listed a single source 0958+56 midway between 0957+561 and NGC3079. Figure 2(b) is reproduced from the recent all-sky survey by Condon and Broderick (1986) made with the NRAO 300-foot telescope at 1400 MHz. Their beamwidth was $\sim$ 10 arcmin, and the point-source response has circular contours. With their greater resolution, it is clear that the "source" listed in our original survey is really two unresolved sources, the stronger being NGC3079 and the weaker being 0957+561. The radio position of 0958+56 listed in Richard's thesis is close to the centroid and is really quite reasonable. It is within 2 arcmin of the 4CP position of the source 4C55.19. Using this position, Caswell and Wills (1967) had identified 4C55.19 with NGC3079. The latter had been detected as a radio source even earlier by Heeschen and Wade (1964). So this area had already been picked over.

The interesting thing is the search procedure that Richard followed with the 300-foot dish. He started with the 4CP position and searched immediately around it but didn't find anything. Then he proceeded to move north. He could have moved south, but he moved north and found a source. He *could* have found NGC3079 then stopped, but he moved north and found 0957+561. The source 0958+56 in our Mark IA survey had a 966-MHz flux density of 1.7 Jy, which puts it well above the survey threshold of 0.7 Jy. If 0957+561 had been a little further away from NGC3079, the two would have been resolved in our survey and we would have identified NGC3079 but not 0957+561. We know now that NGC3079 is much stronger than 0.7 Jy; it should be in our survey and it isn't. 0957+561 is weaker than 0.7 Jy; it shouldn't be in our survey, but it is.

The identification still had to be done. In the summer of 1976, Richard supplied us at Jodrell Bank with a lot of positions. Anne Cohen was doing the identifications using this information. Simultaneously and independently at NRAO, Richard had a summer student, Meg Urry, who also was trying to make identifications using the positions. Anne Cohen had drawn up a preliminary list of identifications in November 1976, which contained the source 0958+56. It was classified as "no identification," but she mentioned "a double-stellar object, magnitude 17.4, 24 arcsec north, not the galaxy of Caswell and Wills." Twenty-four arcsec was a little bit

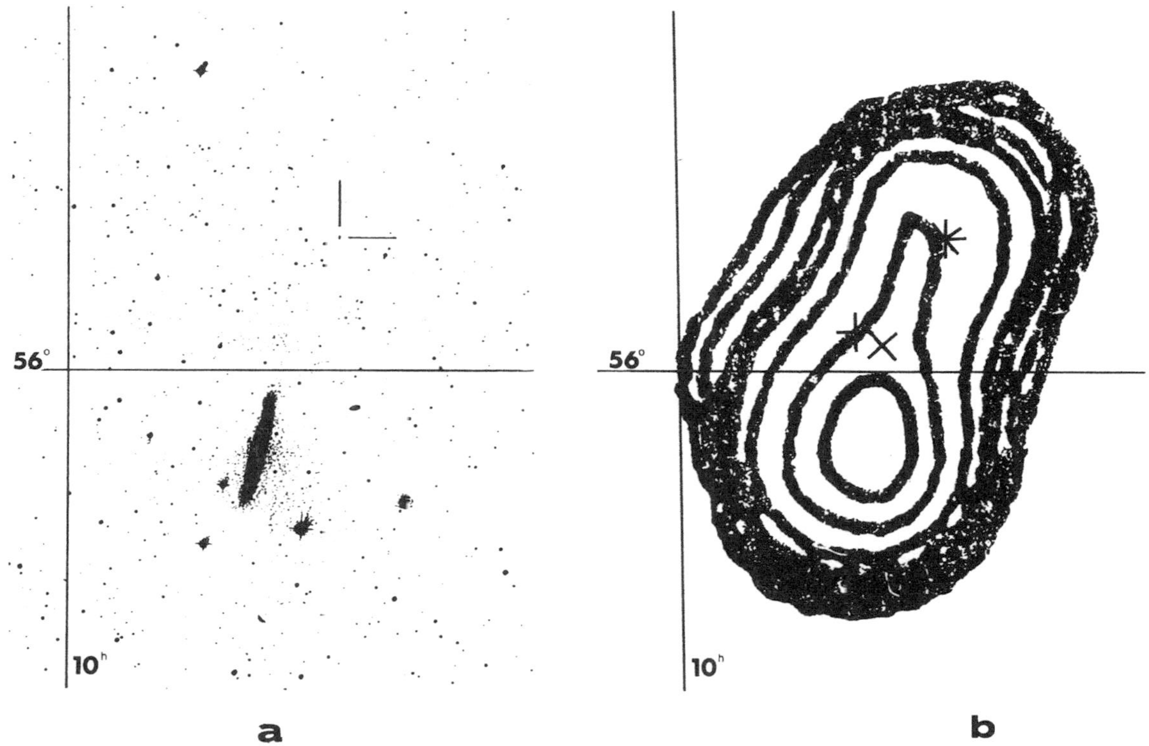

Fig. 2. (a) Field of 0957+561 reproduced from POSS. The "double BSO" is indicated. The prominent galaxy is NGC3079; its separation from 0957+561 is 14 arcmin. Right ascension 10^h and declination 56° indicated. (b) Portion of 1400 MHz Sky Atlas (Condon and Broderick 1986) to the same scale as Fig. 2(a). The peak contour is centered on NGC3079. Indicated are: position of 0957+561 (✱), position of 0958+56 from 966 MHz survey (+), 4CP position of 4C55.19 (×).

far away to make an identification. If you identified everything within 24 arcsec of our positions, you'd end up with a lot of false identifications. However, these were very blue, which made them interesting. Quite separately, Figure 3 shows an extract from Meg Urry's log from the summer of 1976. She draws attention to the pair of objects, one BSO of magnitude 17, five arcseconds north, the second BSO of magnitude 17, ten arcseconds north. Those positions were fortuitously good because the final radio position determined by Richard was 17 arcsec from the nearer BSO,

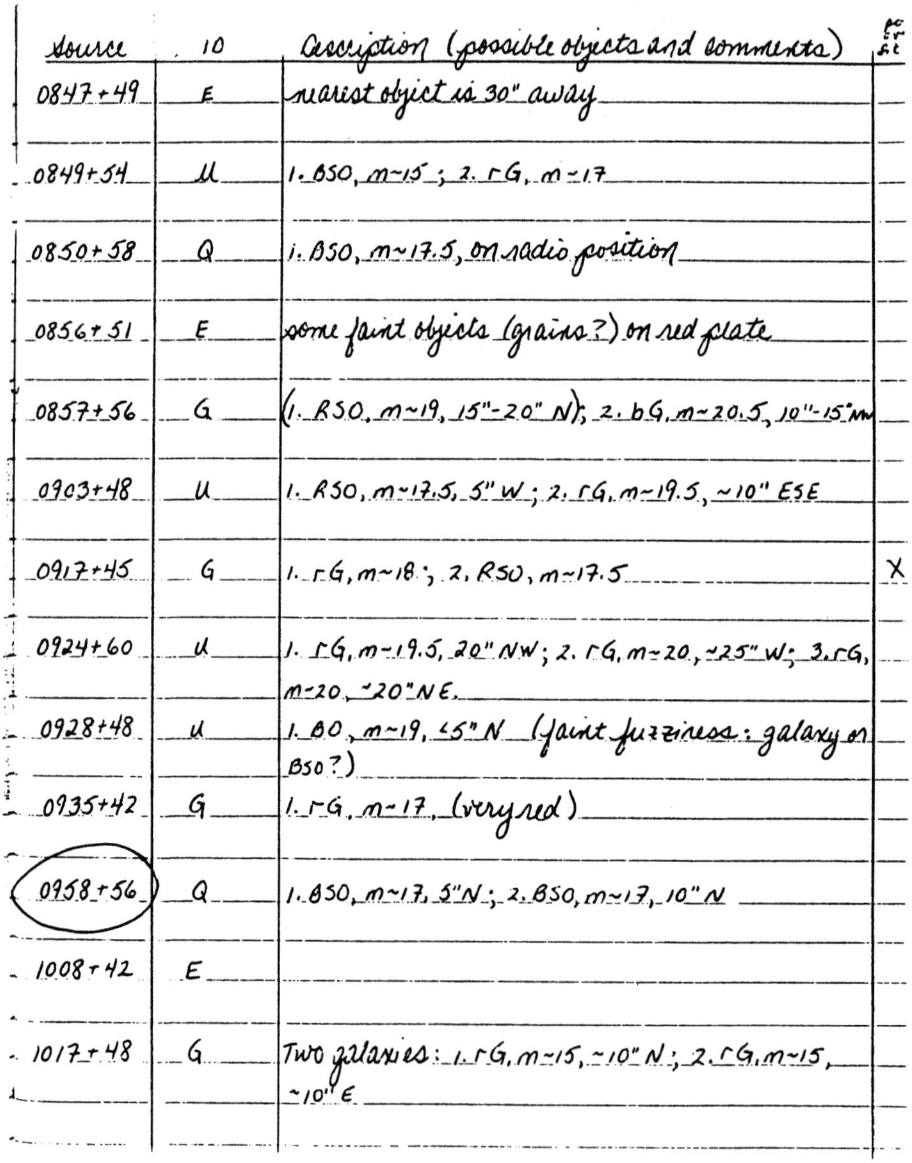

Source	ID	Description (possible objects and comments)	[illegible]
0847+49	E	nearest object is 30" away	
0849+54	U	1. BSO, m~15; 2. rG, m~17	
0850+58	Q	1. BSO, m~17.5, on radio position	
0856+51	E	some faint objects (grains?) on red plate	
0857+56	G	(1. RSO, m~19, 15"-20" N); 2. bG, m~20.5, 10"-15" NW	
0903+48	U	1. RSO, m~17.5, 5" W; 2. rG, m~19.5, ~10" ESE	
0917+45	G	1. rG, m~18; 2. RSO, m~17.5	X
0924+60	U	1. rG, m~19.5, 20" NW; 2. rG, m~20, ~25" W; 3. rG, m~20, ~20" NE.	
0928+48	U	1. BO, m~19, <5" N (faint fuzziness: galaxy or BSO?)	
0935+42	G	1. rG, m~17, (very red)	
0958+56	Q	1. BSO, m~17, 5" N; 2. BSO, m~17, 10" N	
1008+42	E		
1017+48	G	Two galaxies: 1. rG, m~15, ~10" N; 2. rG, m~15, ~10" E	

Fig. 3. Extract from M. Urry's log made at NRAO, summer 1976.

still too far for a confident identification. Richard had several starting positions and we needed him to tell us which one to work with. Anne and Meg were the first to draw attention to the pair of BSO's. However, it's still a long way from drawing attention to an object to having a firm identification of a radio source.

Anne measured the accurate optical position, as published (Porcas *et al.* 1980), on the X–Y machine on 9 June 1977. I personally became aware of it in August 1977. For several years, I'd been trying to obtain optical spectra of the stellar objects in the survey. If we had a blue stellar object, it was almost 100 percent certain that it was a QSO. We had a few red stellar objects too, but the great majority were blue. I'd been working with a number of people on this. Ian Browne worked with me; Alec Boksenberg and Maarten Schmidt took some interest in it, and I got some 200-inch time with them in 1974; Derek and Bev Wills in Texas became interested in taking spectra at McDonald, and I observed with them there in 1976. In 1977 I was preparing an application to Kitt Peak to do more of this work. Figure 4 shows the first record of 0957+561 that I can find in my notes. It is a list of queries to Anne Cohen. I had asked her to give me a list of all the stellar objects in the identification list, and one can see the note adjacent to 0957+561 that reads, "Where did this come from? Magnitude? X,Y?" Note, it's called a double BSO with 6 arcsec separation. I guess this realization came into my consciousness on August 23 in that year. This source was included in an application to Kitt Peak. Inevitably, the application arrived after the September 30 deadline, so we didn't get time at Kitt Peak that year. With a much improved version of the application the next year, we got time.

I was quite excited about this whole thing because, although 17 arcsec away from the radio position wasn't in terribly good agreement (in fact, it was poorer than the criterion for inclusion, which was a search radius of 10–15 arcsec), the two stellar objects were very, very blue. The A object is the bluest object in our whole survey, which is a fortuitous thing, I suppose. My proposal to Kitt Peak was joint with Bob Carswell, with whom I had been working since the earliest days of the spectral work. On our first clear night, we quickly got onto this object. Figure 5(a) shows the first spectrum ever obtained for 0957+561A. It was taken on 29 March 1979 with the IIDS. Anyone familiar with the instrument will know that this was taken at the telescope: we've still got the instrumental profile, which gives the overall shape. After 20 minutes, there are two strong broad emission lines, which were clearly carbon IV and carbon III at redshift 1.4. We couldn't reduce the data fully at the telescope, but the redshift was clearly approximately 1.4. Fine, we had a QSO. Then we moved on to object B. Figure 5(b) is a spectrum of object B obtained a few minutes later. There are two strong emission lines, the

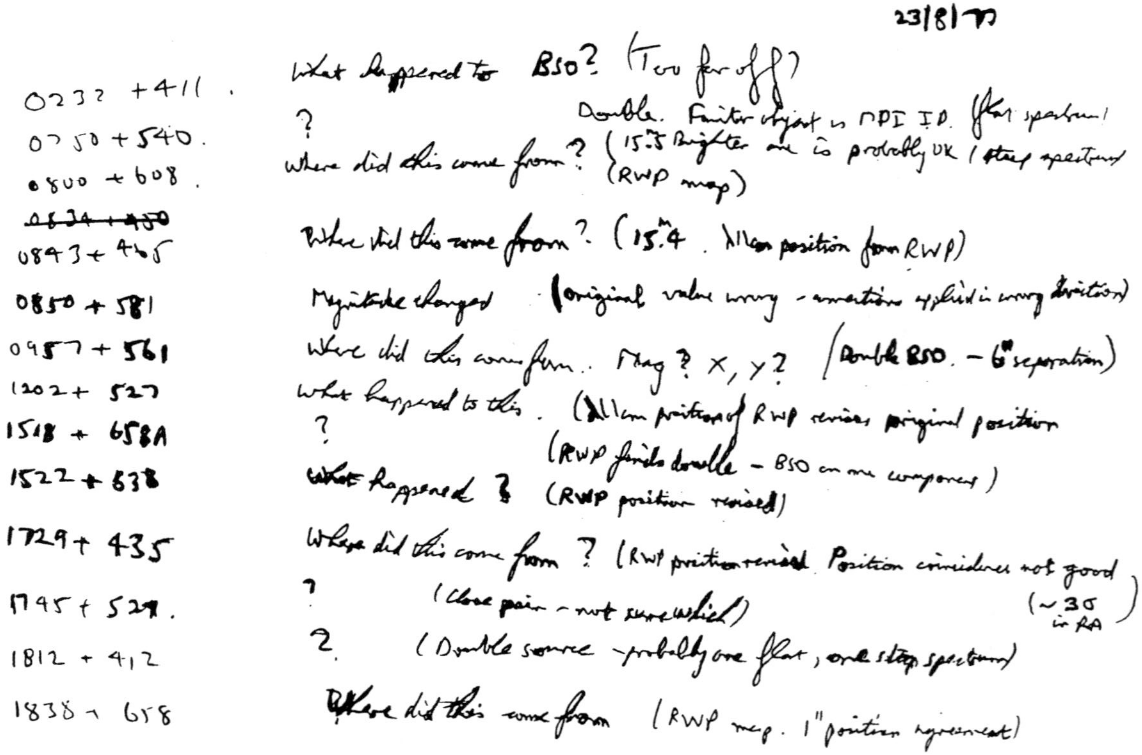

23/8/77

0232 + 411. What happened to BSO? (Too far off?)

0750 + 540. ? Double. Fainter object is MDI I.D. (flat spectrum) (15.5 Brighter one is probably UK / steep spectrum)

0800 + 608. Where did this come from? (RWP map)

~~[illegible]~~

0843 + 465 Where did this come from? (15.4. Allen position from RWP)

0850 + 581 Magnitude changed (original value wrong - correction applied in wrong direction)

0957 + 561 Where did this come from. Mag? X, Y? (Double BSO. - 6" separation)

1202 + 527 What happened to this. (Allen position of RWP revises original position

1518 + 658A ? (RWP finds double - BSO on one component)

1522 + 638 What happened? (RWP position revised)

1729 + 435 Where did this come from? (RWP position revised. Position coincidence not good (~3σ in RA))

1745 + 529. ? (Close pair - not sure which)

1812 + 412 ? (Double source - probably one flat, one steep spectrum)

1838 + 658 Where did this come from (RWP map. 1" position agreement)

Fig. 4. List of queries from Dennis Walsh to Anne Cohen, 23 August 1977. (Partial answers in parentheses.)

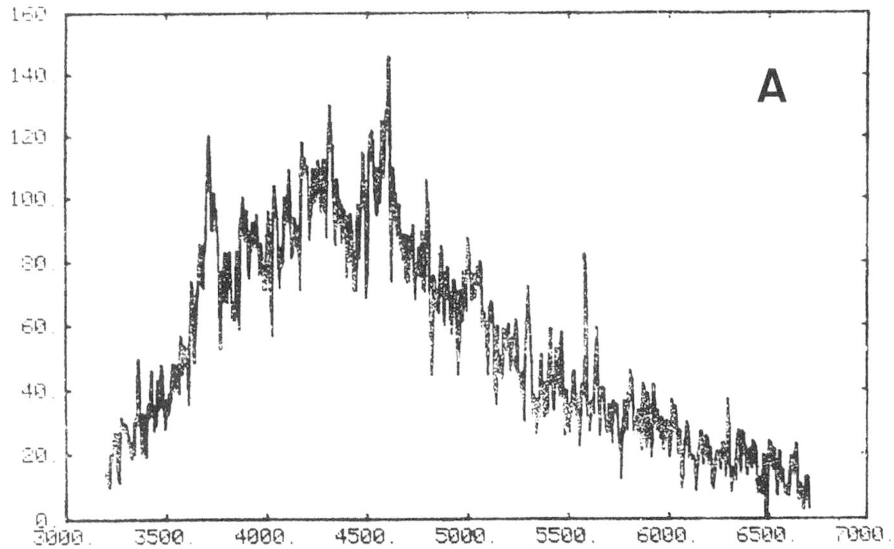

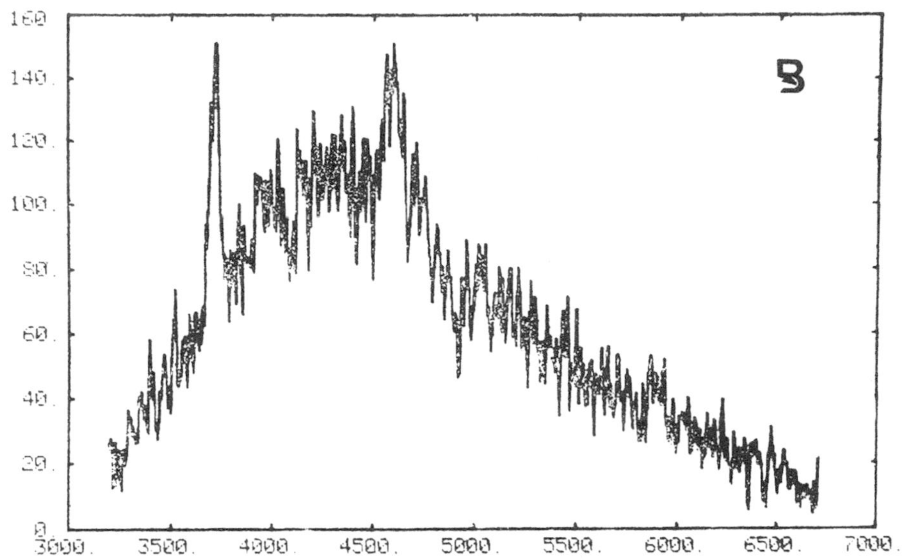

Fig. 5. Upper: First spectrum obtained for 0957+561A at the 2.1-meter telescope, 29 March 1979. Lower: First spectrum obtained for 0957+561B, a few minutes later.

same two emission lines. Same redshift. Clearly, we'd made a mistake and had set on the same object twice. So we went back and made sure. We went back and forth a few times before we were quite certain that these two objects had the same redshift. We did all kinds of tests for light scattering and seeing. There was no contamination of one spectrum by the other. We had two QSO's with the same redshift. We got very excited and thought about this. We did not think right away that it was a gravitational lens. Anybody now finding two objects at similar redshift calls it a gravitational lens, but you didn't do it so readily in those days.

What other observations could we make? Meanwhile, we were steaming along, with the long list of QSO candidates to go through: that's what we'd come to do. While we were exposing these, we were discussing other desirable observations. Clearly, one important observation was a deep image with good resolution. How would we do that? Well, if you go backwards a year to 1978, Ian Browne and I had used the Soviet 6-meter telescope to take deep images of empty fields (empty on the Palomar Sky Survey prints). We had quite a lot of time on the 6-meter but not much success because we had terrible weather and one or two problems with the emulsions. I was keen to do more of it, but with the emulsions available, we were very limited. And so I'd arranged (after a lot of bureaucratic negotiations) for a McMullen camera, an electronographic camera from RGO, to go out to the 6-meter telescope. McMullen himself went there earlier in 1979 to check the interfacing arrangements and so forth, and indeed the camera went out there in the spring. The allocation time at the 6-meter had coincided with the allocation of time at Kitt Peak. I went to Kitt Peak, and Ian Browne had gone out to the 6-meter. We were observing simultaneously, in the same dark run. We wanted a deep exposure, of high resolution; it was perfect. So at 2 a.m. in the morning of 30 March 1979, on our first night at Kitt Peak, I phoned Sir Bernard at Jodrell Bank and said, "Could you get a message through to Ian Browne at the 6-meter to tell them to look at this field?" Sir Bernard sent the telegram off, but they had not received the telegram at the 6-meter when Ian left 10 days later. It never worked, those are the breaks.

Then, the next day, two interesting things happened. First of all, I phoned Derek Wills and told him about it. I'll come back to this later. Second, we talked with Ray Weymann at Kitt Peak. Bob Carswell and Ray Weymann had been long-time collaborators. He was very interested, for two reasons. He had just come up to Kitt Peak because a single night had become available at short notice on the Steward 2.3-meter telescope. I guess that was another lucky break. Ray was doing a study of intermediate dispersion of carbon IV absorption lines in intermediate redshift QSO's, which is exactly the redshift range we were talking about, namely,

1.4. So he was very keen to observe this, and he observed it that night, his one night on the 2.3-meter telescope. It was a superb night, at 1 arcsec seeing, everything was perfect. We got better spectra at the 2.1-meter at low resolution than we got on the previous night. Ray was using an intensified photographic technique, and some time in the early hours of the morning, he phoned to say that he had similar absorption features in both objects. This was the first time that a close pair of QSO's had redshifts so close to each other. Now, we had the first case of two QSO's with common absorption features. Bob went down to the 2.3-meter, while I kept the show running at the 2.1-meter, and came back and said that the absorption features were there. Ray drove back to Tucson as soon as he finished observing.

Meanwhile, Bob and I were still measuring redshifts for a lot more QSO's. The next evening while we were having dinner, a telephone call came from Ray in Tucson that Bob took: he came back and said that Ray had measured the spectra accurately, and the redshifts of the QSO's in the absorption features were as identical as they could be. It looked like we had a gravitational lens. That was the first time that I recall those words being used. I think that it was probably Ray Weymann, who was the first to see the absorption spectra, who used them first. At this point, I had not yet seen the absorption spectra. However, I think that if we had not obtained the absorption spectra, we would have been very hesitant about suggesting we had found the first example of lensing.

Let me move on to other things. In the field, we've got two objects with the same redshifts; maybe there were others. We weren't thinking that maybe there were more than two gravitationally lensed images, though it would be nice to rationalize it in retrospect and say that. We were going to search around. So we took mostly bright stellar objects and took five minute exposures, or something like that. Sure enough, they were all stars. So there were only two objects with this redshift.

That summarizes what we knew when we left Kitt Peak. We quickly drafted our first paper, and the rest of the story is published. I remember well the mixed reaction it received, from immediate conviction to a great wall of disbelief.

One remaining part of the story has not appeared fully in print, although Malcolm Smith (1981) did tell part of it. I said earlier that I'd been doing collaborative work with Derek and Bev Wills; on my way to Kitt Peak in 1979, I passed through Austin to finish off a paper with them, and I talked about what I was going to be doing, and I showed Derek this finding chart of

0957+561 and said, "There are two stellar objects here. What do you think these are?" Derek said, "They're stars." Derek, as some of you may know, had done a very systematic search for pairs of QSO's, he'd looked at about 100 bright QSO's, expecting to find some pairs and found none. So it seemed that there were no pairs, no close pairs. So I made a bet with Derek. I said, "Look, these are very blue, I don't think they're stars," and I wrote on his blackboard, "No QSO's, I pay Derek 25 cents. One QSO, he pays me 25 cents. Two QSO's, he pays me a dollar." It was on the tip of my tongue to say, but I didn't say it, because it sounded facetious, "Two QSO's, same redshift, he pays me $100." When I phoned Derek the next morning and told him what we had found, we laughed and I said, "You owe me a dollar. Suppose I had said to you, 'Two QSO's, same redshift, $100,' would you have taken it?" He said, "Of course." So I lost $99, and I kept a friend. Well, he didn't pay up immediately. Derek and Bev went and looked at them on the 107-inch. A couple of months later, he sent copies of much better spectra than we'd taken, with a silver dollar fastened to them with scotch tape. I kept the silver dollar for many years; unfortunately, I somehow managed to lose it. But it served me very well: I had four teenage sons, none of them particularly interested in science. So when they asked me, "What good is this gravitational lens?" I was able to say, "Well, I made money out of it."

References.

Caswell, J. L., and Wills, D. 1967, *Mon. Not. R. astr. Soc.*, **135**, 231.

Cohen, A. M., Porcas, R. W., Browne, I. W. A., Daintree, E. J., and Walsh, D. 1977, *Mem. R. astr. Soc.*, **84**, 1.

Condon, J. J., and Broderick J. J. 1986, *A 1400 MHz Sky Atlas Covering* $-5° < \delta < +82°$, (Green Bank: NRAO).

Heeschen, D. S., and Wade, C. M. 1964, *A. J.*, **69**, 277.

Porcas, R. W., Urry, C. M., Browne, I. W. A., Cohen, A. M., Daintree, E. J., and Walsh, D. 1980, *Mon. Not. R. astr. Soc.*, **191**, 607.

Smith, M. G. 1981, in *Investigating the Universe*, ed. F. D. Kahn (Dordrecht: Reidel), p. 197.

Walsh, D., Carswell, R. F., and Weymann, R. J. 1979, *Nature*, **279**, 381.

HISTORY OF GRAVITATIONAL LENSES AND THE PHENOMENA THEY PRODUCE

Jeno M. Barnothy
833 Lincoln Street
Evanston, IL 60201

Oliver Lodge was the first to propose in 1919 that light could be focused through a gravitational lens. Following this suggestion, E. B. Frost, director of Yerkes Observatory in 1923, outlined a program to search for lens effects among stars. O. Chwolson (1924) anticipated that the light of distant stars grazing the limb of a foreground star would be seen by an observer aligned with the two stars as a ring around them. It is possible that the luminous arc discovered in 1987 by R. Lynds and V. Petrosian is actually a "Chwolson ring" image of a galaxy. In 1936, Einstein (1936), at the request of R.W. Mandl, made a short calculation and rediscovered the possible existence of the Chwolson ring, now called the "Einstein ring." Einstein concluded that the ring would be of such small size that it could not be resolved by telescopes. He also found that the apparent amplification of the focused star light would become quite considerable when the observer, the lensing star, and the object star are almost exactly aligned. Moreover, he found that the amplification increases proportionally to the square root of the distance to the lensed star and that the ring image splits into two crescents if the sight line through the lensing star bypasses the object. Also in 1936, H.N. Russell wrote a popular paper for *Scientific American* on gravitational lenses, arguing that this phenomenon could provide an experimental test of general relativity so conspicuous that it would be evident even "to the man in the street." In 1937, F. Zwicky (1937) showed that galaxies should act as excellent gravitational lenses and that the detection of gravitational lenses producing visible images would be feasible. Nevertheless, interest in the gravitational lensing phenomenon arose only after Yu. Klimov (1963), S. Liebes (1964), S. Refsdal (1964*a*), and F. Link (1967) independently of each other published detailed calculations of the dioptric properties of gravitational lens images. Bourassa and Kantowski (1975) calculated the dioptric and photographic properties of images produced by distributed mass lenses and showed that such lenses can produce multiple images from the same object. Occasionally, the object itself is seen through the body of the lens. This may be the reason why in the literature it is usually stated that a gravitational lens produces at least three images.

In 1965, at the 119th meeting of the AAS, based on the astonishing similarity of the spectra of quasars and nuclei of Seyfert I galaxies, I proposed (Barnothy 1965) that quasars are not novel superluminous constituents of the Universe, as was generally assumed, but are nuclei of Seyfert I galaxies intensified through the gravitational lens action of foreground galaxies. Moreover, the observed short-term brightness variations of quasars do not prove the extreme smallness of quasars but result from the scanning motion of the optical axis passing across areas of different brightness in the source. The calculations made by J. M. and M. F. Barnothy (1968*a*) showed that if Seyfert galaxy nuclei have an average absolute magnitude of -18 mag, in an FIB (Faraway Information Blurring) universe, we should observe 3,000 quasars of -23 absolute magnitude. As was reported in the November 1983 issue of *Sky & Telescope*, my proposal was met "with amusement by some astronomers, whose snickers were often audible." My idea was also rejected in 1967 by Geoffrey and Margaret Burbidge in their seminal book *Quasi-Stellar Objects* (Burbidge and Burbidge 1967). They argued that the conditions for lensing are inherently implausible; that the lens must be dark objects, otherwise their light would swamp the flux of the gravitationally focused objects; and that the emission lines of QSO's are weak relative to the continuum, whereas in Seyfert I nuclei they are very strong. Perhaps because of this opinion, no efforts were made to find lensed quasars during the next decade. Many astronomers developed theories to explain the extreme luminosity of quasars or the nature of their energy source. Between 1967 and 1976, I published 10 papers pointing out conceptual faults in the proposed explanations.

Interest in gravitational lensing was rekindled in 1979 when Walsh, Carswell, and Weymann (1979) discovered two quasars (0957+561 A and B), separated by $6''.15$, which had identical redshifts ($z = 1.41$), similar emission and absorption lines, and nearly equal magnitudes. They recognized that they were seeing two images of the same object! The lensing galaxy was found by Young *et al.* (1981) to be $0''.8$ north from the southernmost image. The difference in the spectra of A and B may be caused by the light of a giant cD galaxy at $z = 0.036$, probably also obscuring the object itself. For more details, see Burke (1984). Burke initiated, with the collaboration of Hewitt, Turner, Lawrence, Bennett, Langston, and Gunn, a VLA gravitational lens survey. During the nine years since the discovery of 0957+561, 16 further double or triple quasar configurations were discovered, all strongly supporting their interpretation as being images of the same object produced by lensing. Stimulated by the discovery of the first double quasar, astronomers produced so many theoretical and experimental results that space limitations prevent me from giving adequate credit to all of them.

In a Euclidian space, one expects that the number of lensed objects to increase with the 5th power of the distance; however, in a closed hyperspherical space, like that of the FIB universe, the number of objects should follow a sine-square function of distance. Indeed, the histogram of 1640 quasars found with methods other than objective prism follows, with 97% correlation, a sine-square function (Barnothy and Barnothy 1988). This result may explain why Osmer (1982) observed a cutoff in the number of quasars around $z = 4$, suggesting that the first antipole of our position in the Universe is somewhere near there.

The brightness of many quasars varies for several reasons on time scales of days and years: the scanning motion of the optical axis of the lens (Barnothy 1965); in distributed mass lenses through shifting of minilenses into the light path (Canizares 1982); and the explosion of supernovae within the object (Barnothy and Barnothy 1986). In double quasars where the light paths to the images are of different length, the pattern of the brightness variations are the same, but shifted in time. Refsdal (1964*b*) has shown that this phenomenon can be used to determine the value of the Hubble constant as well as the mass of the lens.

The number of double quasars among the more than 3,000 known quasars is much smaller than what one would expect. The reason for this is that a very massive lens is needed to produce a resolvable double image. The Hubble Space Telescope may help to discover more. Moreover, light coming from a distant object is scattered during its passage through the gravitational field of galaxies, causing a blurring of the image (Barnothy and Forro 1944). This effect may lead to an overlap of the images.

Many astrophysical phenomena discussed in the literature could be produced by gravitational lenses. I give several examples.

(1) The gravitationally lensed images are brighter because of the apparent enlargement of the object area. Velocities between elements of a radio source, being vector quantities, will be seen to be similarly increased, occasionally appearing as superluminal velocities (Barnothy 1982).

(2) Neutron stars are excellent minilenses. In neutron star binaries, one star can act as the ocular lens while the other acts as the objective lens, thus forming a "telescope." Whenever an observer is close to the sight line, the observer will, during every revolution, see two light flashes lasting up to 100 milliseconds duration (Barnothy and Barnothy 1968*b*). At the 145th meeting of the AAS in Bloomington (Barnothy 1975), I demonstrated

the operation of this neutron binary star telescope, on a model in which the neutron star lenses were simulated by properly molded plastic lenses. I offered this model as an alternative explanation of pulsars.

(3) The intensification of an element of a source through a gravitational lens is inversely proportional to the distance of the element from the lens axis. If the area emitting the emission line is farther from the axis than the area emitting the continuum, then in a lensed object, we see the quasar, but not its emission lines, a phenomenon called a BL Lac object. This explanation of the BL Lac phenomenon occurred independently to Ostriker and Vietri (1985) and to Barnothy and Barnothy (1986). The latter pointed out that a significant correlation between equivalent width of emission lines and continuum, the Baldwin effect, should occur only in QSO samples containing mainly lensed objects.

(4) The FIB universe is described by the static solution of the Friedmann equations. It is a closed hypersphere of spherical geometry; its radius and entropy fluctuates within narrow limits. One can say that the distributed mass of this universe acts as a "topological" gravitational lens, focusing the light rays emitted by a source, in its successive antipoles on the hypersphere. The images of the source that are produced in even-numbered antipoles overlap the source. This "topological lens" explains the surprising brightness of quasars with $z > 3.5$, in the vicinity of the first antipole of the Galaxy. According to our calculations, the latter is about $z = 5.6$ (see Barnothy and Barnothy 1988). Taking our Galaxy as the light source, the spectrum of its starlight is fairly well matched by that of a blackbody of 5800° K. After two turns around the Universe, the light of the Galaxy will return, redshifted to $z = 1886$, as a microwave background radiation (MBR) of 3.06° K. The image of the Galaxy is now blurred into a halo of 45 kpc size, centered upon the Galaxy. This phenomenon explains not only the great isotropy of MBR when seen by an observer positioned at the center of the halo, but also why the primary cosmic radiation is not cut off above 5×10^{17} ev energy through its interactions with MBR, where its absorption length would have been about 500 kpc. This explanation of the MBR as a gravitational lens focused light of our Galaxy, obviates its commonly accepted origin: a singularity preceeding the Big Bang. The topological lens should also focus the light emitted by the Galaxy after one turn around the Universe. We could hence expect to observe a peak in the Wien region of MBR. Indeed, recently Matsumoto *et al.* (1988) have observed a second peak in the submillimeter range.

(5) The topological gravitational lens also focuses the neutrino radiation of galaxies, leading to matter production where matter is present. We cannot go here into more details of this process.

References.

Barnothy, J. M. 1965, *A. J.*, **70**, 666.

__________. 1982, in *Extragalactic Radio Sources, IAU Symp. 97*, eds. D. S. Heeschen and C. M. Wade (Reidel, Dordrecht), p. 463.

Barnothy, J. M., and Barnothy, M. F. 1968*a*, *Science*, **162**, 348.

__________. 1968*b*, *A. J.*, **73**, 164.

__________. 1975, *Bull. AAS*, **7**, 263.

__________. 1986, *A. J.*, **91**, 755.

__________. 1988, *Comments Astrophys.*, **13(4)**, in press.

Barnothy, J. M., and Forro, M. 1944, *Csillagaszati Lapok*, **7**, 65.

Bourassa, R. R., and Kantowski, R. 1975, *Ap. J.*, **195**, 13.

Burbidge, G., and Burbidge, M. 1967, *Quasi-Stellar Objects* (W.H. Friedman and Co.).

Burke, B. 1984, *Comments Astrophys.*, **10**, 75.

Canizares, C. R. 1982, *Ap. J.*, **263**, 508.

Chwolson, O. 1924, *Astr. Nachrichten*, **221**, 329.

Einstein, A. 1936, *Science*, **84**, 506.

Klimov, Yu. G. 1963, *Soviet Phys.*, **8**, 119.

Liebes, S. 1964, *Phys. Rev.*, **133**, B835.

Link, F. 1967, *Bull. Astr. Inst. Czech.*, **18**, 215.

Lodge, O. J. 1919, *Nature*, **104**, 354.

Matsumoto, T., Akiba, M., and Murakami, H. 1988, *Ap. J.*, **332**, 575.

Osmer, P. 1982, *Ap. J.*, **253**, 28.

Ostriker, J. P., and Vietri, H. 1985, *Nature*, **318**, 446.

Refsdal, S. 1964*a*, *M. N. R. A. S.*, **128**, 295.

__________. 1964*b*, *M. N. R. A. S.*, **128**, 307.

Russell, H. N. 1937, *Scientific American*, **156**, 76.

Walsh, D., Carswell, R. F., and Weymann, R. J. 1979, *Nature*, **279**, 381.

Young, P., Gunn, J. E., Kristian, J., Oke, J. B., and Westphal, J. 1981, *Ap. J.*, **244**, 736.

Zwicky, F. 1937, *Phys. Rev. Lett.*, **51**, 290 and 679.

III. Optics and Models

THE VERSATILE ELLIPTICAL GRAVITATIONAL LENS

Ramesh Narayan and Scott Grossman
Steward Observatory, University of Arizona
Tucson, AZ 85721 USA

Introduction. At the present time, about ten cases of gravitational lensing are known where either a distant quasar is multiply imaged or a background galaxy is imaged into an arc or a ring. Many of these cases have been modeled in detail, but no single model has been applied to all the cases. The purpose of this paper is to demonstrate that a simple elliptical lens can reproduce approximately the positions of the images and their magnifications/shapes in most of the observed cases. We are not attempting to compete with the more highly parametrized models, which do lead to better agreement with the observations, but we wish merely to point out that an uncomplicated generic model can reproduce the observations surprisingly well.

The Model. The examples shown in Figures 1 to 3 are based on the lens model described by equations (2) and (3) below. However, we must emphasize that the results are quite insensitive to the details of the model — any lens that is approximately isothermal and roughly elliptical in projection will have qualitatively similar properties.

The lensing equation is given by

$$\vec{r}_S = \vec{r}_I - \vec{\alpha}\,(\vec{r}_I) = \vec{r}_I - \vec{\nabla}\psi\,(\vec{r}_I)\,, \tag{1}$$

where $\vec{r}_S$ and $\vec{r}_I$ are vectors that measure angular coordinates in the source and image planes. $\vec{\alpha}\,(\vec{r}_I)$ is the reduced bending angle due to the lens and $\psi\,(\vec{r}_I)$ is an effective two-dimensional lens potential (e.g., Blandford and Narayan 1986). We model $\psi\,(\vec{r}_I)$ as the sum of a circularly symmetric component (monopole) and an elliptical distortion (quadrupole). Following Grossman and Narayan (1988), we assume the following form for the monopole term,

$$\begin{aligned}\psi_0\,(r_I) &= \alpha_0 r_c \left[\tfrac{3}{8}\left(\tfrac{r_I}{r_c}\right)^2 - \tfrac{1}{32}\left(\tfrac{r_I}{r_c}\right)^4 + \tfrac{21}{32}\right], && r_I \le r_c,\\ &= \alpha_0 r_c \left[\tfrac{r_I}{r_c} - \tfrac{3}{8}\ln\left(\tfrac{r_I}{r_c}\right)\right], && r_I \ge r_c.\end{aligned} \tag{2}$$

This corresponds to a surface density that varies quadratically for $r_I < r_c$, matching smoothly to a $1/r_I$ dependence ("isothermal") for $r_I > r_c$. The reduced bending angle is α_0 for $r_I \gg r_c$. The quadrupole term is assigned the form

$$\psi_2\left(r_I, \theta_I\right) = \frac{\epsilon \alpha_0 {r_I}^2}{2\left(r_c + r_I\right)} \cos 2\theta_I, \tag{3}$$

where ϵ measures the degree of ellipticity of the lens. The velocity dispersion of the lens and the angular diameter distances to the lens and the source are absorbed into α_0, which describes the angular scale of the deflection. Apart from this scale, the lens model has two free parameters: the core radius, r_c/α_0, and the ellipticity, ϵ.

Caustics and Critical Lines. In order to understand the imaging characteristics of a lens, it is useful to study the structure of its *caustics* and *critical lines.* Caustics are lines in the source plane that demarcate regions of different image multiplicities. For typical parameters, the above lens model has two caustics, dividing the source plane into regions of multiplicity 1, 3 and 5 respectively (see the Source Planes in Figs. 1 and 2). The change in multiplicity when the source crosses a caustic occurs through the creation/destruction of a pair of merged images on a critical line in the image plane.

If a point source were to cross a caustic from the lower multiplicity side, two new images would appear as a merged pair on the associated critical line and would then separate from

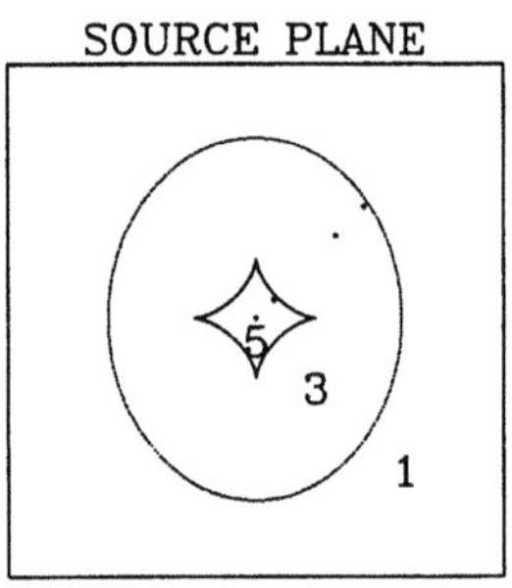

Fig. 1. The inner and outer lines in the Source Plane are the tangential and radial caustics, which separate regions of image-multiplicity 1, 3, and 5. The inner and outer lines in the Image Panels are the radial and tangential critical lines. Panels a-d show image geometries corresponding to the four source positions (dots in the Source Plane), starting from the center. The angular scales in the Source Plane and Image Panels are not the same. The bottom panels show observed multiply imaged quasars that resemble the model configurations. A-D are quasar images and G represents the center of the lensing galaxy (if observed).

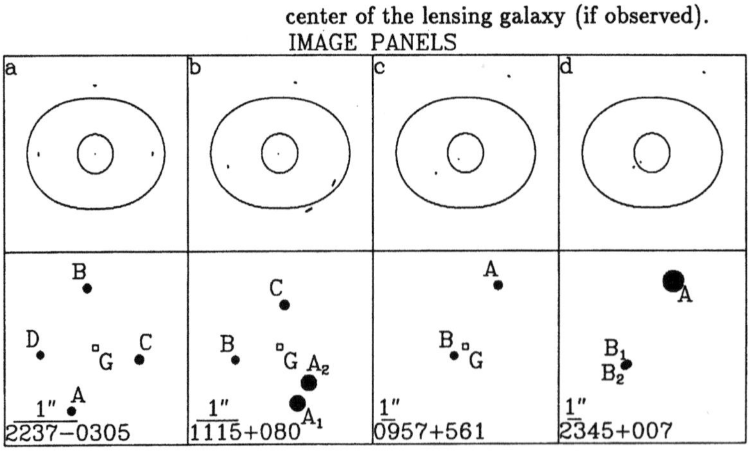

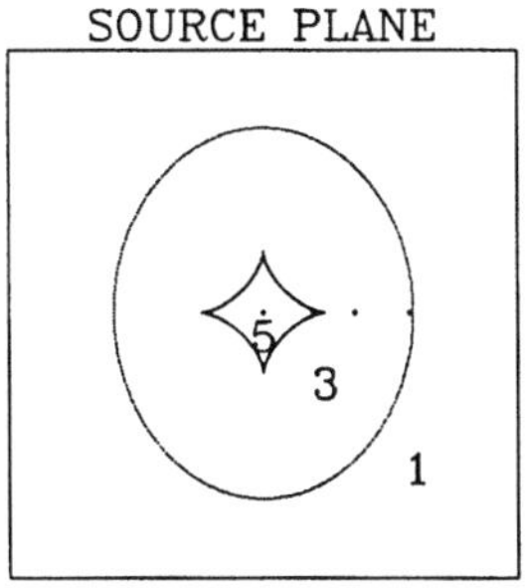

Fig. 2. As in Fig. 1, but for a different sequence of source positions. When the source is just inside the cusp, three images merge on the tangential critical line (panel b); compare with panel b of Fig. 1 where the source is inside a fold and only two images merge. The radial and tangential nature of the two caustics can be distinguished by observing the direction of merger of the images in panels b and d. Here and in Fig. 1, the lensing potential is assumed to have the form given by eqs. (2) and (2) with $r_c/\alpha_0 = 0.2$, $\epsilon = 0.1$.

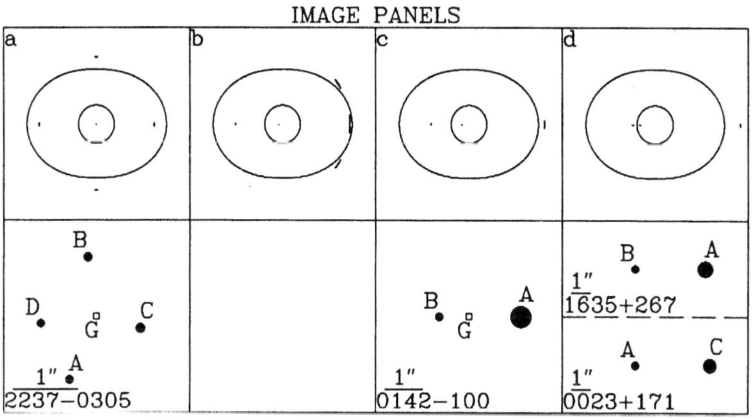

each other. For a generic elliptical lens model such as the one considered here, the direction of separation tends to be approximately radial with respect to the lens center for the 1-to-3 caustic and approximately tangential for the 3-to-5 caustic (see the Image Panels in Figs. 1 and 2). Therefore we label the former as the *radial caustic* and the latter as the *tangential caustic*, with similar labels for the associated critical lines. The magnification of the pair of images varies inversely as their angular separation and so close pairs tend to be very bright (see Figs. 1 and 2, where the areas of the images are proportional to the magnifications). In the case of an extended source, the images are elongated along their direction of separation, again by an amount that is inversely proportional to the image separation. Thus, the brightest multiply imaged quasars and the most extensive arcs tend to be associated with caustics and critical lines; hence the importance of these lines.

The above description is for a generic caustic, that is to say, a *fold caustic*. In addition, there can be singular points on the caustic line that represent so-called *cusp caustics*. These lead to three bright merging images when the source is inside the cusp and a single bright image when the source is on the outside. In the lens model considered here, the tangential caustic has four cusps and the radial caustic has none.

Roughly speaking, the nature of the radial caustic is determined primarily by the monopole term in the potential, while the tangential caustic is influenced strongly by the quadrupole. In the next two sections we compare our model predictions with the observations and we show that the tangential caustic plays an important role. This means that the ellipticity of the lens is a crucial ingredient of the model. Indeed, a circularly symmetric lens cannot explain any of the lensed quasars with more than three images, and cannot produce the kinds of arcs that have been observed (Grossman and Narayan 1988).

Image Geometries. Figures 1 and 2 show sequences of source positions and their associated image configurations for a typical galaxylike lens ($r_c/\alpha_0 = 0.2, \epsilon = 0.1$). Beneath the image panels are shown observed cases of multiply imaged quasars with similar morphologies. [See Blandford and Kochanek (1988) and references therein for a description of the observations up to 1986; more recent observations are cited below.] The main point of Figures 1 and 2 is that a fairly generic sequence of source positions reproduces the broad features of nearly all the multiply imaged quasars discovered so far.

The image panels a-d in Figure 1 correspond to the four source positions in the Source Plane, starting from the center. Panels a and b are cases with five images but, in practice, only four images will be seen since the central image tends to be much fainter than the others. In panel a, the four images occur in the vicinity of the tangential critical line and are symmetrically placed around the center of the lens. The lensed quasar, Q2237-0305, has an image arrangement that is a slightly distorted version of this configuration (Schneider *et al.* 1988). The case shown in panel b corresponds to a source position close to one of the folds in the tangential caustic. Two of the four images approach each other and are significantly brighter than the other two. The resemblance of this configuration to the "triple" quasar, Q1115+080, is striking. In panel c the source has crossed the tangential caustic into the three-image region, and therefore we have two bright images and a faint third image. The two bright images have roughly equal magnifications and are noncolinear with respect to the lens center and nonequidistant from it. This is very similar to the geometry of the original "double" quasar, Q0957+561. Finally, panel d shows the case when the source approaches the radial caustic. Two of the images approach each other across the radial critical line; this case resembles the image configuration observed in Q2345+007 (Nieto *et al.* 1988).

Figure 2 shows another sequence of four source positions, this time passing through one of the cusps in the tangential caustic. Panel b corresponds to a source located just inside the cusp and has three very bright tangentially merging images. We are a little surprised that no multiply imaged quasar has been found with this configuration. However, as we explain in the next section, this is the typical geometry for arcs from extended sources and several of these have been observed. In panel c, one of the outer images is significantly brighter than the other because of the influence of the cusp caustic. It is probable that Q0142-100, with the largest A to B magnification ratio found so far, belongs to this case. It is even conceivable that Q0142-100 corresponds to panel b, with the triple image being unresolved. High resolution observations of this object are of interest. Finally, we classify Q1635+267 and Q0023+171 under panel d of Figure 2; however, since these objects consist merely of two images with no lensing galaxy seen, they could equally well belong to panel c of Figure 2 or panels c or d of Figure 1.

Our simple elliptical lens model cannot explain the multiply imaged quasar, Q2016+112. However, two lensing galaxies, C and D, have been seen in this case and it is therefore not surprising that a model with a single lens fails.

Arcs. When the background source is not a pointlike quasar but an extended galaxy, the images can be observably distorted into highly elongated "arcs", particularly when the source lies on a caustic (Grossman and Narayan 1988). Figure 3 shows several configurations that could lead to arcs with a typical lens with parameters appropriate to a cluster of galaxies ($r_c/\alpha_0 = 0.55$, $\epsilon = 0.1$). The most dramatic arcs are associated with cusps, where three images of the source merge tangentially. The case in panel a is similar to the arcs discovered in the clusters A370 and Cl2244-02 (Soucail *et al.* 1987, Lynds and Petrosian 1987), while the case in panel b, with a counter image, resembles the arc in A963 (Lavery and Henry 1988). The central image in b

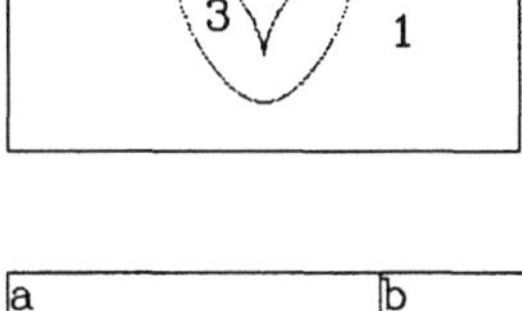

Fig. 3. Elongated images of extended sources (filled circles) for four locations in the Source Plane. Image Panels a and b correspond to sources on cusps; three images merge tangentially on the tangential critical line to form a large arc. Panels c and d correspond to folds. In panel c, a small arc is produced by the tangential merger of two images, and in panel d a radially elongated image is formed. The lens potential corresponds to $r_c/\alpha_0 = 0.55$ (appropriate for a galaxy cluster, which is likely to have a less concentrated core relative to its bending than a galaxy) and $\epsilon = 0.1$.

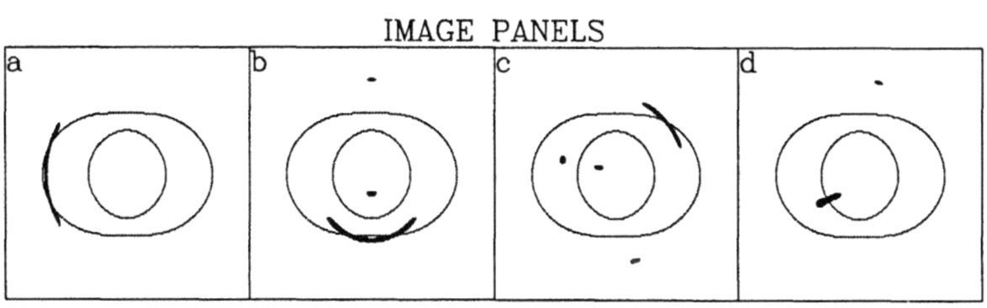

will be reasonably bright if the cluster has a nonsingular core but will be very weak if a cD galaxy is present, as is the case in all three clusters. Panel c shows a smaller arc associated with a fold section of the tangential caustic and panel d shows a radially elongated image corresponding to the radial caustic. Images similar to c have been seen, but nothing like d seems to have been found so far; this may not be surprising since the radial arcs are much less spectacular than their tangential counterparts and will be suppressed even further if a cD galaxy is present in the cluster core. An elliptical lens is also remarkably successful in reproducing the ring image of the radio source MG1131+0456 (Hewitt *et al.* 1988, Kochanek *et al.* 1988).

Conclusion. We illustrated in this paper the variety of image geometries that is possible with a simple elliptical "isothermal" lens, organizing the possibilities with the aid of the caustics and critical lines. We then showed that all the multiply imaged quasars discovered so far (with the exception of Q2016+112), as well as the luminous arcs, can be understood in terms of this model. We find this very reassuring since the most likely gravitational lenses that we know of, namely, galaxies and clusters of galaxies, are roughly isothermal and elliptical in projection. Therefore, as far as *image morphologies* are concerned, the observations to date are consistent with our view of the universe at cosmological distances. We must caution, however, that we have not considered the *magnitudes* of the image separations (described by α_0). It is well known that there is a problem in this case — the observed image separations in the multiply imaged quasars tend to be somewhat larger than those expected for isolated galaxylike lenses, implying the presence of significant amounts of dark matter in these lenses.

As we mentioned earlier, the results presented in this paper do not depend critically on the lens model. Other models of an elliptical lens have been studied, e.g., Kovner (1987), Blandford and Kochanek (1987).

Acknowledgments. RN first learned about the properties of elliptical lenses and the importance of caustics from R. Nityananda, to whom he is indebted. Some of the ideas presented in this paper developed from discussions with R. Blandford, C. Kochanek. and I. Kovner, and we thank them. This work was supported by NSF grant AST86-11121.

References.

Blandford, R. D., and Kochanek, C. S. 1987, *Ap. J.*, **321**, 658.

Blandford, R., and Kochanek, C. S. 1988, *Proc. Thirteenth Jerusalem Winter School on Dark*

Matter in the Universe, ed. T. Piran, in press.

Blandford, R., and Narayan, R. 1986, *Ap. J.*, **310**, 568.

Grossman, S. A., and Narayan, R. 1988, *Ap. J. (Letters)*, **324**, L37.

Hewitt, J. N., Turner, E. L., Schneider, D. P., Burke, B. F., Langston, G. I., and Lawrence, C. R. 1988, *Nature*, **333**, 537.

Kochanek, C. S., Blandford, R. D., Lawrence, C. R., and Narayan, R. 1988, *M. N. R. A. S.*, in press.

Kovner, I. 1987, *Ap. J.*, **312**, 22.

Lavery, R. J., and Henry, J. P. 1988, *Ap. J. (Letters)*, **329**, L21.

Lynds, R., and Petrosian, V. 1987, *Bull. AAS* **18**, 1014.

Nieto, J.-L., Roques, S., Llebaria, A., Vanderriest, Ch., Lelièvre, G., di Serego Alighieri, S., Macchetto, F. D., and Perryman, M. A. C. 1988, *Ap. J.*, **325**, 644.

Schneider, D. P., Turner, E. L., Gunn, J. E., Hewitt, J. N., Schmidt, M., and Lawrence, C. R. 1988, *A. J.*, **95**, 1619.

Soucail, G., Fort, B., Mellier, Y., and Picat, J. P. 1987, *Astr. Ap.*, **172**, L14.

GRAVITATIONAL LENS OPTICS

R. D. Blandford
Harvard-Smithsonian Center for Astrophysics
60 Garden St., Cambridge, MA 02138, USA
and
Theoretical Astrophysics
130-33, Caltech, Pasadena, CA 91125, USA

Recently, interest in gravitational lenses has expanded to include the study of the images of extended sources such as galaxies as well as of those of unresolved point sources like quasars. The giant luminous arcs associated with the rich clusters Abell 370, Cl2244-02, and Abell 963, together with the ring radio source MG1131+0456, are now widely believed to be gravitationally imaged. Some simple geometrical ways to understand the formation of extended images are described in this article. It is explained how the image shape is controlled by the location of the source relative to the caustic surfaces formed by the lens. Multiply-imaged extended sources generally create larger magnifications and furnish more information about the gravitational potential of the lens than do multiply-imaged point sources. A simple iterative method for deriving the source structure is described and some preliminary results for MG1131+0456 are exhibited. Future observational prospects are briefly discussed.

Introduction. Although the elementary theory of gravitational lenses was developed (e.g., Refsdal 1964, Bourassa and Kantowski 1975) prior to the discovery of Q0957+561AB (Walsh, this volume), it took the stimulus of having actual examples to model to force astronomers and physicists to apply notions long familiar from telescope design to cosmic refractors. Gravitational lenses do, in a loose sense, realize Zwicky's dream of a cluster-sized telescope capable of magnifying the most distant galaxies. This partly explains the fascination that they hold for astronomers. However, the "telescopes" themselves are also interesting, and by studying them, we can probe dark matter on both galactic and cosmological scales.

Professor Burke was one of the first astronomers to react to the discovery of Q0957+561, and he has had a leading role in the subsequent development of the field (for reasons upon which I can only speculate, Fig. 1). It is characteristic of him that the subject that we are discussing here is not some half-forgotten, 30 year old scientific triumph but instead a topic that has become even more exciting since this meeting was announced. I am delighted to be here to help refuel his enthusiasm.

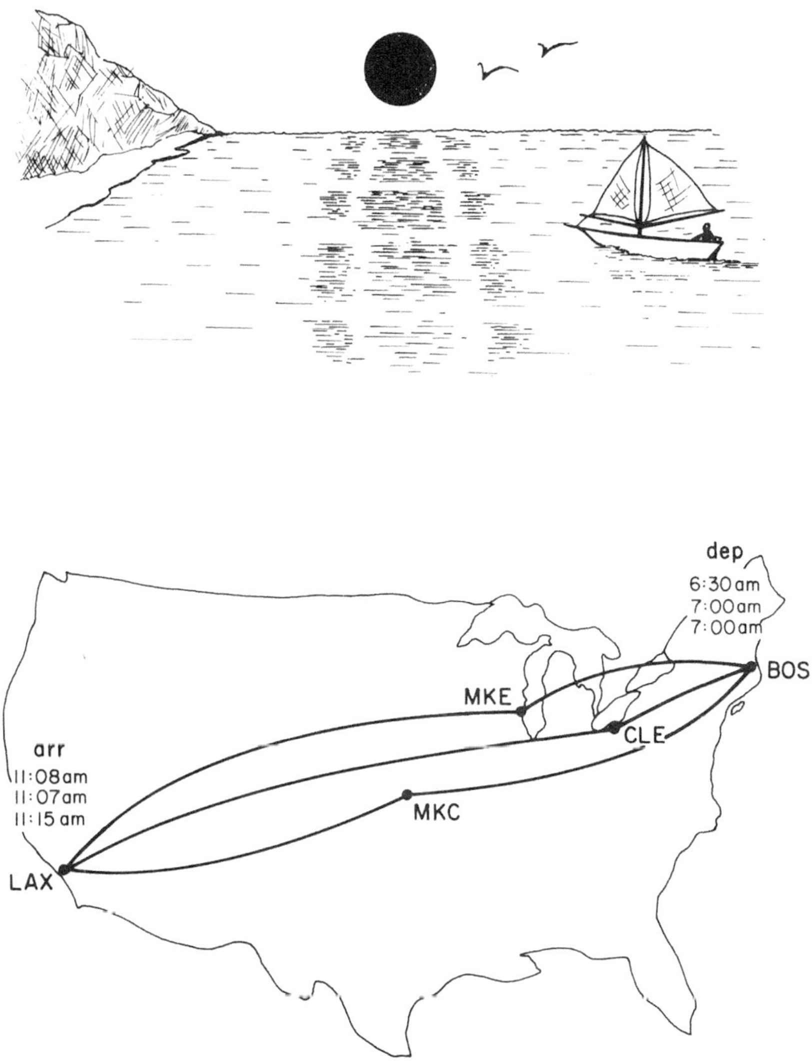

Fig. 1. Two possible explanations for B. Burke's interest in gravitational lenses.

A major reason for this recent surge in interest in gravitational lensing has been the detailed observations of the giant arcs (Petrosian, Soucail, this volume). These arcs are probably the images of background galaxies observed through giant clusters and differ from the earlier examples of lensing because the sources are significantly extended. In this contribution, I want to outline some ways of thinking about the geometrical optics of gravitational lenses by emphasizing the imaging of extended sources. The accompanying article by Grossman and Narayan will emphasize the imaging of point sources. I shall not cite many original references; more comprehensive bibliographies can be found in recent reviews by Canizares (1987), Blandford and Kochanek (1987), and Blandford *et al.* (1988).

Elliptical Lenses. A gravitational lens differs in two distinct ways from a simple lens in which rays from a point source converge to or diverge from a point. Firstly, the deflection caused by a gravitational lens varies nonlinearly with the displacement of a ray from the optic axis. This characteristic allows multiple images to form. It is most convenient to consider rays extending backward in time from the observer (i.e., us) to the plane of the source. The angles that these rays make with the optic axis (which we define by extending a line from us through the center of the lens) lie in the *image plane.* The angles that would be made by source points at some fixed distance from us in the absence of the lens are suitable coordinates for the *source plane.* The lens maps the image plane onto the source plane.

We can illustrate this scenario using a nonlinear, though circularly symmetric, lens. If the source is on the optic axis, it can be connected to us by a ray along the optic axis and also by rays that make a finite angle with the optic axis and form a circular "Einstein" ring in the image plane as shown in Figure 2 (e.g., Saslaw *et al.* 1985). Now, let us introduce the idea of a *ribbon* of rays, coplanar with the optic axis. As the lens is circularly symmetric, a ribbon will remain coplanar with the optic axis after passing by the lens and will intersect the source plane in short lines or *spokes* (Fig. 3). These spokes will pass through a common center. A ribbon

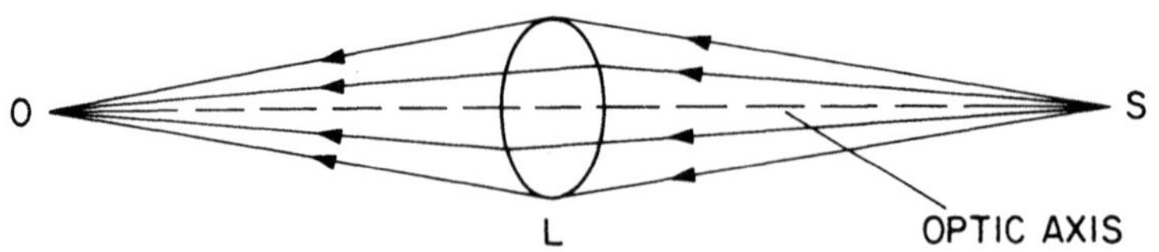

Fig. 2. Formation of an Einstein ring by a circularly symmetric lens. S is a source lying in the source plane on the optic axis. L is the lens and O the observer.

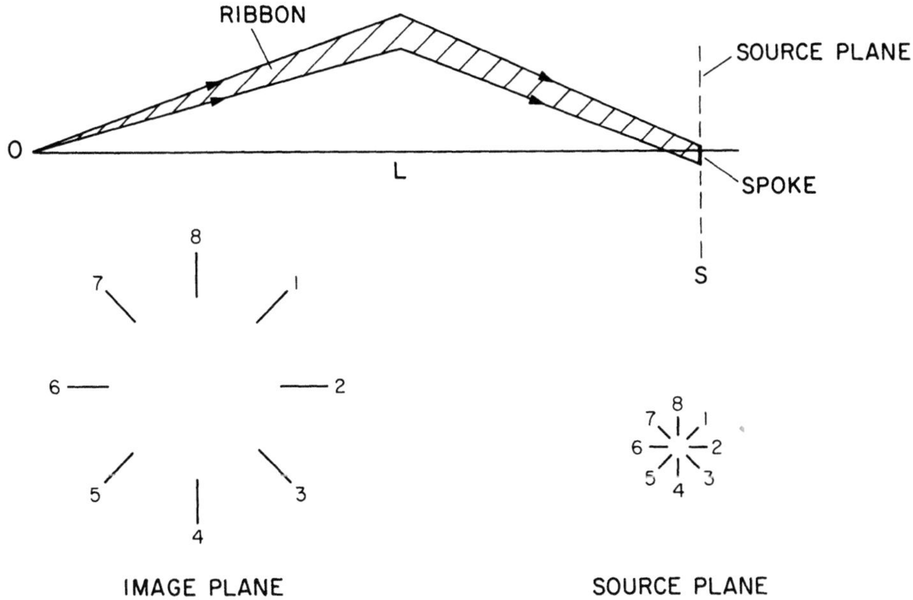

Fig. 3. Ribbons of rays propagating backwards from us past a circularly symmetric lens. The ribbons, labeled 1–8, intersect the source plane in radial "spokes," which should be extended as necessary.

maps a line of points lying on a spoke in the source plane onto the image plane, in the present case of a circular lens, onto a short radial segment. A source point near the optic axis lies on two spokes on opposite sides of the optic axis, and it will therefore be mapped onto two diametrically opposed image points each close to the Einstein ring, one just inside the ring, the other just outside it. (There will also be a third image close to the center of the Einstein ring, either absorbed or highly deamplified, but that need not concern us here.)

The second way in which gravitational lenses differ crucially from simple lenses is that they are usually noncircular. The optics change radically when the circular symmetry is broken. The simplest example involves making the lens astigmatic by adding a small elliptical perturbation. The ribbons of rays emanating from us are then deflected tangentially by the nonradial gravitational field in the lens and the spokes no longer intersect at a point (Fig. 4). Source points in the vicinity of the optic axis are now located at the intersections of either two or four spokes and therefore map onto either two or four image points near the Einstein ring.

Caustics. The family of spokes in the source plane defines an envelope, which is the locus of points where two intersecting spokes become parallel. This is the *caustic*. Source points on the

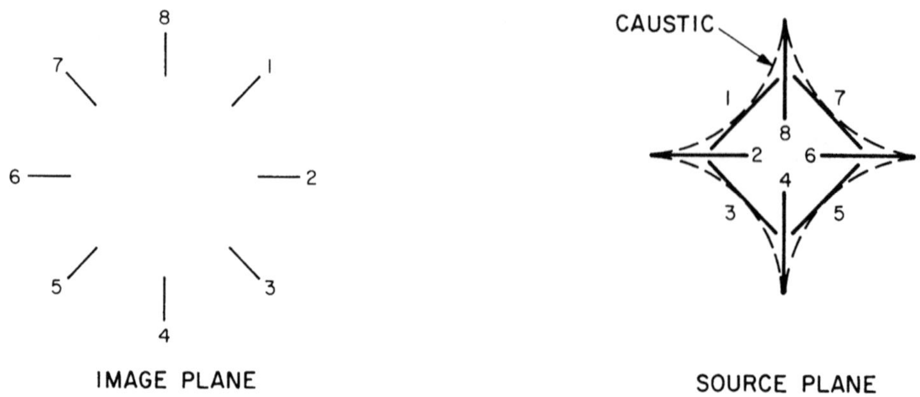

Fig. 4. As for Fig. 3 except that a small elliptical perturbation is introduced into the lens. The spokes in the source plane no longer pass through a common point and instead combine to form an astroid-shaped caustic envelope.

caustic map onto a curve that is a slight deformation of the original Einstein ring, in the image plane. This is called a *critical curve*. This way of dividing up the source plane using caustic curves and the image plane using the associated critical curves is quite general. A spoke will be mapped onto an almost straight line segment in the image plane where the distance of the image from the critical curve is directly proportional to the distance of the source point from the caustic measured along the spoke. A point just inside the caustic will lie on two nearly parallel spokes and will be mapped onto a pair of image points separated tangentially almost parallel to the critical curve.

This caustic curve is an example of the most elementary form of *catastrophe* called the *fold*. Folds can merge at *cusps*, which are higher order catastrophes where the number of neighboring images produced by a source point changes from one to three, rather than zero to two as for a fold. In the present example of a small elliptical perturbation, the caustic is a four-cusped astroid curve and the critical curve is an ellipse of small eccentricity. There is also a much larger caustic curve associated with radially merging images, which need not concern us here. Sources within the astroid produce a total of four images near the critical curve (again ignoring the central image); sources outside the caustic produce two images.

It is useful to extend the idea of a caustic curve in the source plane to the more fundamental notion of a two-dimensional *caustic surface*, which is the envelope surface formed by all the rays emanating from the observer and traveling backward in time (e.g., Blandford and Narayan 1986, Padmanabhan and Subramanian 1988; Fig. 5). These rays touch the caustic surface tangentially. What this implies is that if we isolate one ray leaving us along some

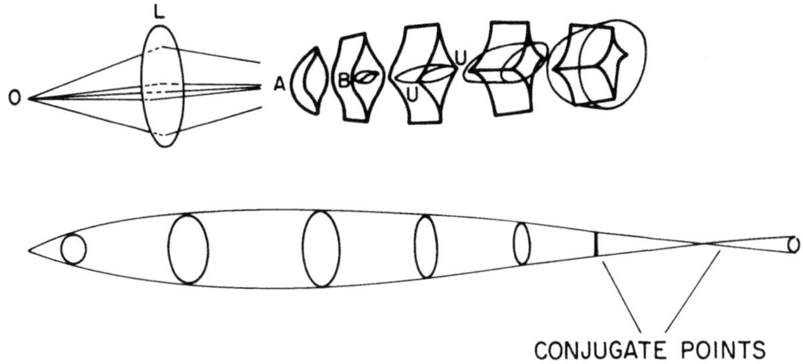

Fig. 5. a) Caustic surface formed by an elliptical lens. Two fold surfaces A,B form behind the lens, terminated by cusp lines. They touch at a higher order catastrophe known as a hyperbolic umbilic, U. The source must usually be more distant than the hyperbolic umbilic for the lens to be regarded as nearly circular and for image formation to be described by the tangent method outlined in the text. b) Evolution of a ray congruence propagating backward in time from us. The cross-section of the rays becomes a line at two foci "conjugate" to us. The rays and the line are tangent to the caustic surface at these two points. Tangentially merging images are produced when the source lies close to the first of these points.

direction and then surround it with a pencil or *congruence* of adjacent rays, then the point of tangency is a point where nearby rays are brought to a focus. If, initially, the pencil has a circular cross-section, it will be deformed into an ellipse as it propagates backwards through the universe past the lens. As the caustic surface is approached, the eccentricity of the ellipse increases until at the point of tangency, the cross-section has flattened to a line lying tangent to the surface. This line, when extended at either end, becomes a spoke. In a conventional cosmology, all rays will be focussed eventually, although not of course always at a redshift of interest (Hawking and Ellis 1973).

Let there be a source point at a given distance behind the lens and just inside the caustic. In this case, there will be two particular rays passing through the source point and connecting to us, one having just touched the caustic surface, the other just about to touch it. These two rays define an "osculating" plane. Pencils associated with both rays will have their cross-sections elongated nearly parallel to the caustic and perpendicular to the osculating plane. These two rays will clearly form images of opposite parity.

If we ignore the expansion of the universe (and to include it only involves a minor modification), then we can understand the magnification of the image by reversing the direction of the congruences, i.e., by considering pencils leaving a point source and being focussed down to a similar highly eccentric ellipse at the observer. The observed flux (power received per unit cross-sectional area) necessarily diverges as the source approaches the caustic. It is an elementary exercise to show that the flux magnification of each image asymptotically increases inversely with the square root of the distance of the source from the caustic surface. Furthermore, the observed angular separation of the two images also increases as the square root of the source distance from the caustic surface. This result is a generic property of fold caustics.

We can now use the caustics to locate approximately the images of a point source located behind a nearly circular lens (Blandford and Kovner 1988). We first construct the spokes by drawing all the tangents to the caustic curve passing through the source (Fig. 6). These spokes are then extended until they intersect the critical curve where the images will form. (The sense of the extension can be determined by inspection.) Using this construction, we can verify that source points just inside cusps create three close images. These three images merge to form a single image when the source lies outside the cusp.

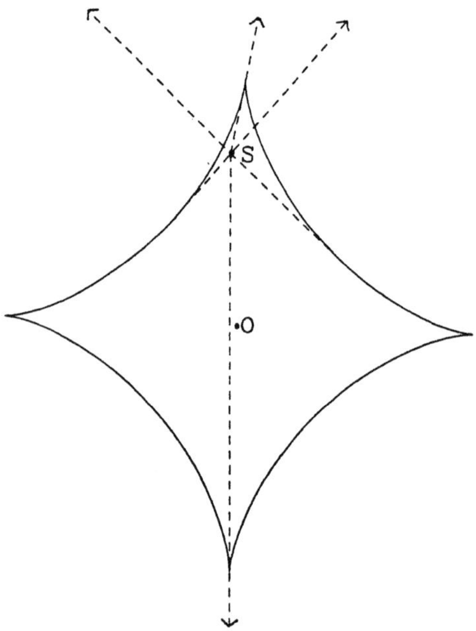

Fig. 6. Directions of images formed by a point source S located within an astroid caustic. The directions are tangent to the caustic.

Catastrophes like the fold and the cusp are very important because their properties are universal in the sense that as a source approaches a singularity, the magnifications, time lags, etc., satisfy asymptotic scaling laws. This fact has been exploited by Kovner and Paczyński (1988) who have pointed out that when a source has three images located near a cusp, the magnifications of any two are in inverse proportion to their separations from the third. One way to demonstrate this is to use the above construction to model a cusp by a semi-cubical parabola and to solve for the gradients of the three tangents connecting the source to the caustic curve. This produces a cubic equation, the roots of which give the directions of the images. The magnifications of the images are controlled by the expansions in the tangential direction, as the radial magnification of the images is essentially constant. These in turn are inversely proportional to the gradients of the cubic at the roots. The above identity then follows from elementary algebra. A similar identity connects the relative time delays.

Extended Sources. The reason for introducing the above geometrical construction is that it is particularly useful for understanding the imaging of arcs and rings. Here the sources are believed to be galaxies that have angular sizes that are not necessarily small compared with the size of the caustics formed by the cluster gravitational lenses. Now, an extended source is just a collection of points that, if located near the tangential caustic, will be imaged near the outer critical curve. We can determine the approximate angular extent of the resulting arcs by drawing the common tangents to the caustics and the source and extending them outwards to the critical curve (Fig. 7).

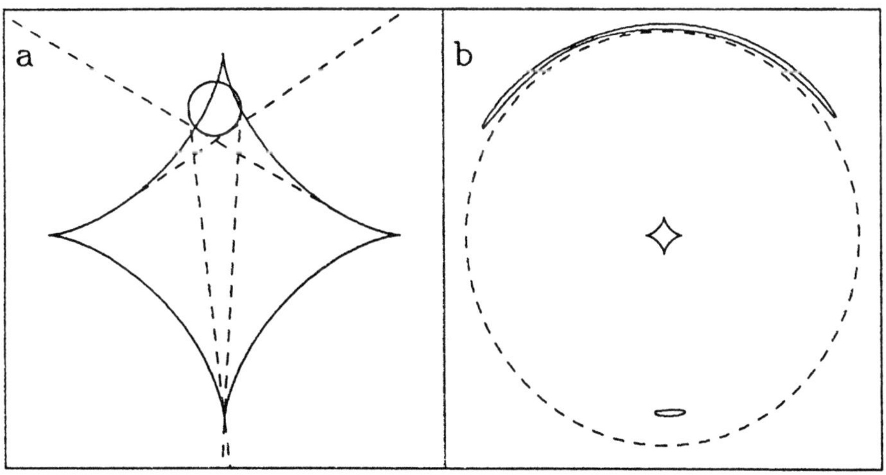

Fig. 7. Arc formation by an extended source. a) In this example, a circular source is located close to a cusp and the directions of the images of the source points are bounded by drawing the common tangents to the caustic and the source. b) Extent of the arc images near to the Einstein ring.

We can immediately see a difference between arcs formed by a circularly symmetric lens and arcs formed by an elliptical lens. In the former case, the caustic degenerates to a point and if the source is displaced from and does not cover the optic axis, then two arcs of equal angular length will be created. However, in the latter case, a source located near a cusp in the caustic will create two arcs of very different length. This is the typical case as demonstrated graphically in simulations by Grossman and Narayan (1988) and Kovner (1988a).

Another important point can be seen from this construction. The source need not be located inside the cusp for a long arc image to be formed. In fact, the source can lie entirely outside the cusp and still create a prominent arc. In a typical case, the source will straddle the caustic and part of it will be singly imaged in the prominent arc with the remainder being triply imaged.

Arcs and Rings. With the discovery of the first arcs (Soucail *et al.* 1987, Lynds and Petrosian 1987), several astronomers (notably Paczyński 1987) suggested that they could be gravitationally imaged galaxies. Several factors have combined to make this explanation more attractive. In the case of Abell 370, the arc is now reported to have a plausible redshift and the color of a distant spiral galaxy (Soucail, this volume). The velocity dispersion of the cluster has been revised upward to 1700 km s^{-1} to make its estimated surface density large enough to allow multiple imaging. The report of additional arcs associated with this cluster (distinguishable by their blue color) also supports the lens explanation as has the realization from both simulations and theoretical analysis that extended arcs will not, in general, be accompanied by equally extended counter-arcs (Grossman and Narayan 1988, Kovner 1988a, Narasimha and Chitre 1988).

It is unclear how common arcs will turn out to be. The ones that have been reported appear to be located in clusters that have higher central surface densities than normal. If we model the cluster by an isothermal sphere with small core radius and one-dimensional velocity dispersion σ, then the radius of the Einstein ring, which is also roughly the radius of curvature of the arcs, will be given by $\sim 4\pi\sigma^2 D_{LS}/c^2 D_{OS}$ with D_{LS} and D_{OS} being the angular diameter distances that connect the lens to the source and the observer to the source, respectively. For a typical distance ratio of ~ 0.5, a velocity dispersion of ~ 1200 km s^{-1} is necessary to create a $\sim 20''$ radius arc. This is at the high end of the distribution of local cluster velocity dispersions, and we generally expect high redshift clusters to be less concentrated than local clusters.

The radio ring MG1131+0456 was proposed by its discoverers (Hewitt *et al.* 1988) to be a smaller scale version of the same phenomenon with the cluster lens replaced by an individual galaxy with velocity dispersion of ~ 250 km s^{-1}. The observed ring is nearly complete, suggesting that the source covers a large fraction of the caustic. Again, simulations, described below, encourage confidence in this interpretation.

Iterative Inversion of Source Structure. Images of extended sources should furnish more information about the shape of the lensing potential than images of point sources. With a point source, each image point provides just one datum, the relative magnification. However, extended images contain redundant information when parts of the source are multiply imaged. This provides a powerful set of constraints on models of the source. Furthermore, when the image actually surrounds the lens, as in MG1131+0456, there is an extensive probe of the shape of the lensing potential. Motivated by the discovery of the arcs and greatly encouraged by the subsequent discovery of a ring, Kochanek *et al.* (1988) devised an iterative technique for inverting the images of extended sources. In this method, we adopt a relatively simple lens model specified by a few (typically no more than 10) adjustable parameters. For the case of MG1131+0456, where a single elliptical galaxy is believed responsible, an elliptical potential suffices, but if, by contrast, there were known to be two galaxies present in the lens, then a superposition of two circular potentials (Kochanek and Apostalakis 1988) should provide a good starting model. The image and source planes are then divided into pixels. For a given set of lens parameters, the intensities of the image pixels can be mapped, many to one, onto the source pixels. This allows us to define a figure of merit for this particular lens model by computing the dispersion in the separate intensities (which will be zero in a true model). The lens parameters are then varied so as to minimize this dispersion. Multifrequency and polarization data can be combined using this procedure.

Application of this method to MG1131+0456 leads to a quite plausible source model, a core-jet radio structure quite common among extragalactic sources and a galaxylike potential (see Fig. 8). However, it also leads to a quite interesting prediction. The best-fitting lens model is elongated roughly perpendicular to the radio ring. However, optical observations of the source region show a galaxy elongated parallel to the elongation of the radio ring (Turner, private communication). Further observations are needed to determine if this preliminary conclusion will have to be revised, and also to refine the lens model.

Observational Ramifications. I will conclude by making some general remarks about the usefulness of arcs and rings. As we have seen, they do indeed provide magnified images of

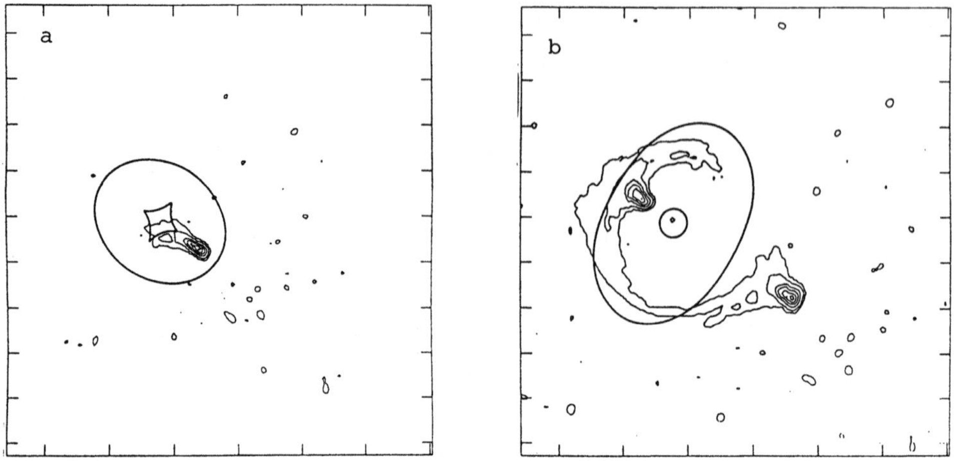

Fig. 8. a) 15 GHz source structure for MG1131+0456 derived using inversion procedure outlined in the text. b) Associated image that is quite similar to the observed image (Hewitt *et al.*, this volume).

cosmologically distant sources. A *typical* linear magnification near cusps by a factor of up to 10 (i.e., roughly the reciprocal of the ellipticity of the arc or equivalently the ratio of the radius of the Einstein ring to the size of the caustic) can be expected. This compares favorably with increases in angular resolution anticipated from the Hubble Space Telescope (HST) and orbiting VLBI, e.g., Radioastron.

If the type of inversion procedure outlined above turns out to be generally useful, then there is the prospect of learning more about the internal structure of young galaxies and compact radio sources. This is because, when an extended source covers a caustic curve, the *maximum effective* magnification will be much larger than the typical magnification. For a fold caustic, the magnification can be increased by a further factor equal to roughly the ratio of the radius of the Einstein ring to the telescope angular resolution, if we assume that the lens is smooth and that the source has structure on the associated linear scale. For optical observations using HST, this will amount to linear structure on the scale of ~ 1 pc. For VLBI observations, the typical linear size resolvable can be as small as 10 AU. In effect, we can look at a small slice of a cosmologically distant radio source with the same resolution that we can look at the center of our galaxy! (Despite these impressively small dimensions, geometrical optics is still valid and the "telescope" is not diffraction-limited.)

Now the problem is to recognize the location of the critical curve as it bisects a galaxy or radio source image. One technique that conceivably might be useful at optical wavelengths is to use the pixel to pixel fluctuations. Tonry and Schneider (1988) have demonstrated that the $\sqrt{N}$ fluctuations in the number of stars can, in principle, be used to count the stars in galaxies out to ~ 20 Mpc and hence furnish distance estimates. By use of HST, for which the maximum cluster-assisted resolution will be $\sim 10^{-4}$ arcsec, pixel to pixel fluctuations should be seen comparable with those that should be observable in galaxies a thousand times closer. It is obviously very difficult to calculate the fluctuation level *a priori*, as stars are clustered and nonstellar contributions like HII regions may be important. Furthermore, the background from the sky and the instrument may make this difficult to accomplish in practice. Nevertheless, empirical comparisons of the fluctuation levels seen across the arcs with those of similar galaxies at different distances may provide important insights into the evolution of galaxies.

So far, none of the arcs have proven to be detectable radio sources, which is not too surprising because the estimated probability of finding a radio source behind the tangential caustic of a cluster is quite small. However, the probability that a galaxy lies between us and a distant radio source is much larger, and it is reasonable to expect that several more ring sources like MG1131+0456 remain to be discovered. There are $\sim 2 \times 10^8$ bright galaxies out to a redshift $z \sim 0.5$ and their Einstein rings cover a fraction $\sim 10^{-3}$ of the sky. The probability of ring formation will be roughly the fraction of sky covered by their tangential caustic curves, $\sim 10^{-4}$, for a typical ellipticity of ~ 0.1. This suggests that several ring images remain to be found. Magnifications of $\sim 10^4$ can be found close to the critical lines in these ring images and the VLBA should have adequate dynamic range to map them so as to be able to identify the critical curves. A demonstration that the critical lines should be quite noticeable given adequate linear resolution and dynamic range is given in Fig. 9. Of course the greatest interest surrounds the compact cores of quasars, but here we cannot hope that any of them lie on a caustic and so do not expect to see any more than a typical magnification of ~ 10. Extra-rapid superluminal expansion may be seen.

As we mentioned in the introduction, the second potential use of arcs and rings is that they provide a diagnostic for the shape of the cluster potential well. In most cases, we expect this to be basically isothermal in shape, that is to say the two-dimensional potential increases roughly linearly with radius outside the core. Clusters are neither expected nor observed to be spherical, and if their formation under gravitational collapse is in any way similar to the formation of galaxies, quadrupolar deviations from circular symmetry should dominate. Kovner (1988b) has used the observations currently available and tried to fit the shapes of the three best observed arcs. He has found that quadrupolar lenses are adequate to reproduce the observations.

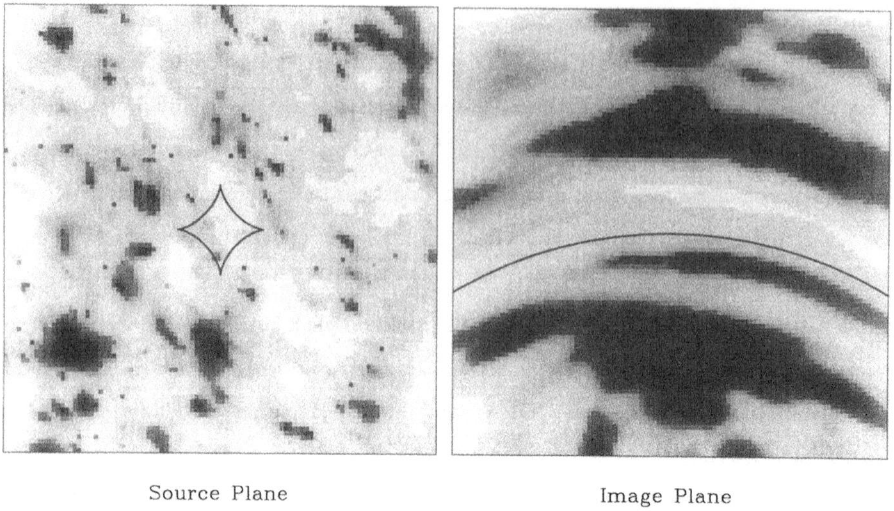

Fig. 9. a) Simulated extended source structure produced by the superposition of 500 uncorrelated Gaussians in the source plane. The astroid caustic of an associated elliptical lens is shown as a bold curve. b) Derived image structure, smoothed to represent the finite angular resolution of the observation. Note the strong shear of roughly circular source features near to the critical curve (also bold).

He has also concluded that in A963, the ellipticity of the underlying mass surface density distribution is ~ 0.5 smaller than that of the central cD galaxy and that in 2242-02, the center of mass is displaced from the center of light, an important warning for those who use galaxies to model the distribution of dark matter in a cluster. In addition, it was necessary to assume that shorter unseen counter-arcs were hidden behind cluster galaxies.

Unfortunately, as brought out by the simulations of Grossman and Narayan (1988), the corrugation introduced into the potential by individual cluster galaxies can strongly influence the length and curvature of the arcs. This will also probably cause problems with Kovner and Paczyński's ingenious proposal to observe the relative magnifications and time lags in a triply imaged supernova occurring in the source galaxy. This is because in using the universal form of a cusp singularity, they are implicitly assuming that the potential is smooth on the scale of the separation of the images. As the arrival time surface is necessarily very flat near a cusp, very small galaxy-induced fluctuations can be highly significant. The magnitude of the time delay fluctuation will be $\sim 10GM/c^3 \sim 10$ days for a galaxy mass $\sim 10^{10}\ M_\odot$, which is comparable for the estimated delay $\sim 17(D_t/1h^{-1}\mathrm{Gpc})(\Delta x/5'')^4(20''/\theta)^2$ days computed by Kovner and Paczyński assuming a pure cusp model. Here Δx is the image separation, θ is the arc radius, and D_t is the usual combination of angular diameter distances, $(1+z_L)D_{OL}D_{OS}/D_{LS}$. We will, of course, have to be lucky to see a supernova in any given arc (Linder, Schneider and Wagoner

1988). Conversely, if galactic perturbations are important, then it will not be possible to use the magnitude of any delays measured to give a useful measurement of the Hubble constant.

The prospects for performing cosmography seem a little more promising for the radio rings, because here we expect to find variable, compact cores, and the lenses may be isolated galaxies that should be more accurately described by elliptical potentials than clusters. Detailed modelling can verify this hypothesis. In addition, there are less likely to be complications arising with microlensing because radio sources are significantly larger than supernovae.

Acknowledgments. I thank my collaborators, C. Kochanek, I. Kovner, C. Lawrence, R. Narayan, S. Phinney, and G. Soucail for many discussions on these and related topics. I am indebted to J. Moran for helpful editorial suggestions. Support by the National Science Foundation under NSF grant AST86-15325, the John Simon Guggenheim Foundation, and the Harvard Smithsonian Center for Astrophysics is gratefully acknowledged.

References.

Blandford, R. D. and Kochanek, C. S. 1987, in *Dark Matter in the Universe, Proceedings of 4th Jerusalem Winter School for Theoretical Physics*, ed. J. Bahcall, T. Piran, and S. Weinberg (World Scientific: Singapore), p. 133.

Blandford, R. D., Kochanek, C. S., Kovner, I., and Narayan, R. 1988, *Science*, (in press).

Blandford, R. D. and Kovner, I. 1988, *Phys. Rev. A.*, (in press).

Blandford, R. D. and Narayan, R. 1986, *Ap. J.*, **310**, 568.

Bourassa, R. R. and Kantowski, R. 1975, *Ap. J.*, **195**, 13.

Canizares, C. R. 1987, in *Gravitational Lenses as Tools in Observational Cosmology, Proc. IAU Symposium No. 124*, ed. A. Hewitt, G. R. Burbidge and L.-G. Fang (Reidel, Dordrecht, Holland).

Grossman, S. and Narayan, R. 1988, *Ap. J. (Letters)*, **324**, L37.

Hawking, S. W. and Ellis, G. F. R. 1973, *The Large Scale Structure of Spacetime* (Cambridge University Press, Cambridge).

Hewitt, J. N., Turner, E. L., Schneider, D. P., Burke, B. F., Langston, G. I., and Lawrence, C. R. 1988, *Nature*, **333**, 537.

Kochanek, C. S. and Apostalakis, J. 1988, *M. N. R. A. S.*, (in press).

Kochanek, C. S., Blandford, R. D., Lawrence, C. R., and Narayan, R. 1988, *M. N. R. A. S.*, (in press).

Kovner, I. 1988a, in *The Post-Recombination Universe*, NATO ASI, ed. N. Kaiser and A. N. Lasenby (in press).

Kovner, I. 1988b, (preprint).

Kovner, I. and Paczyński, B. 1988, (preprint).

Linder, E. V., Schneider, P., and Wagoner, R. V. 1988, *Ap. J.*, **324**, 786.

Lynds, R. and Petrosian, V. 1987, *Bull. AAS*, **18**, 1014.

Narasimha, D. and Chitre, S. M. 1988, (preprint).

Paczyński, B. 1987, *Nature*, **325**, 572.

Padmanabhan, T. and Subramanian, K. 1988, *M. N. R. A. S.*, **233**, 265.

Refsdal, S. 1964, *M. N. R. A. S.*, **128**, 295.

Saslaw, W. C., Narasimha, D., and Chitre, S. M. 1985, *Ap. J.*, **292**, 348.

Soucail, G., Fort, B., Mellier, Y., and Picat, J. P. 1987, *Astr. Ap.*, **172**, 414.

Tonry, J. and Schneider, D. P. 1988, *A. J.*, **96**, 807.

GRAVITATIONAL LENSING OF EXTENDED SOURCES

S. M. Chitre
Institute of Astronomy, Cambridge, England, and
Tata Institute of Fundamental Research, Bombay, India

D. Narasimha
Tata Institute of Fundamental Research, Bombay, India

The emphasis in gravitational lens theory has hitherto been largely on the imaging of a point source by either a point mass or by an extended deflector. It was only recently that the importance of the study of gravitational lensing of finite-size sources was recognized for its possible role in generating interesting geometrical configurations in the extragalactic domain. For example, the lensing of an extended source such as an elliptical galaxy by a transparent lens in the form of a cluster of galaxies could simulate the giant arcs observed in the cores of some clusters. Rays from an extragalactic radio source with a central compact region and two hot spots, for example, could be intercepted by a giant foreground galaxy to give rise, with suitable alignment, to a nearly circular ringlike feature with multiple bright nuclei. It is also possible to contruct a scenario in which a high-redshift ($z \gtrsim 1$) quasar with a jet may be lensed by a nearby galaxy ($z \lesssim 0.05$) along the line of sight to generate an image configuration that shows the quasar to have an apparent luminous connection with the outer contours of the foreground galaxy. Even for the confirmed high-redshift gravitational lens system 2016+112, the predictions of models of the extended emission-line regions turn out to be qualitatively different from those based on a point source model. Our model calculations are shown in Figures 1 to 5.

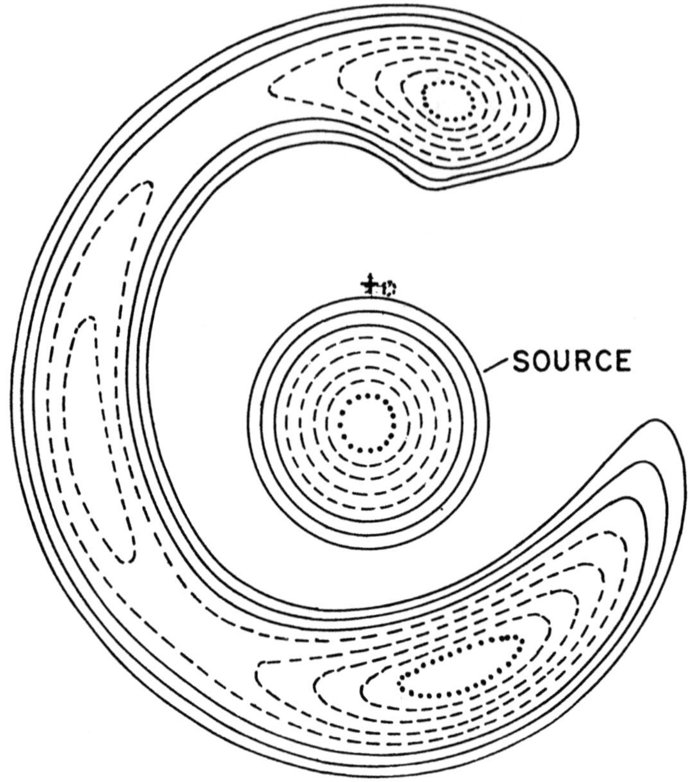

Fig. 1. A schematic illustration of a background source consisting of a bright nucleus and surrounding fuzz being lensed by an intervening object to product an almost ringlike image configuration with bright spots, shown on an arbitrary scale. The source is indicated by circular contours of varying intensity, and the cross denotes the center of the deflector (galaxy or galaxy cluster).

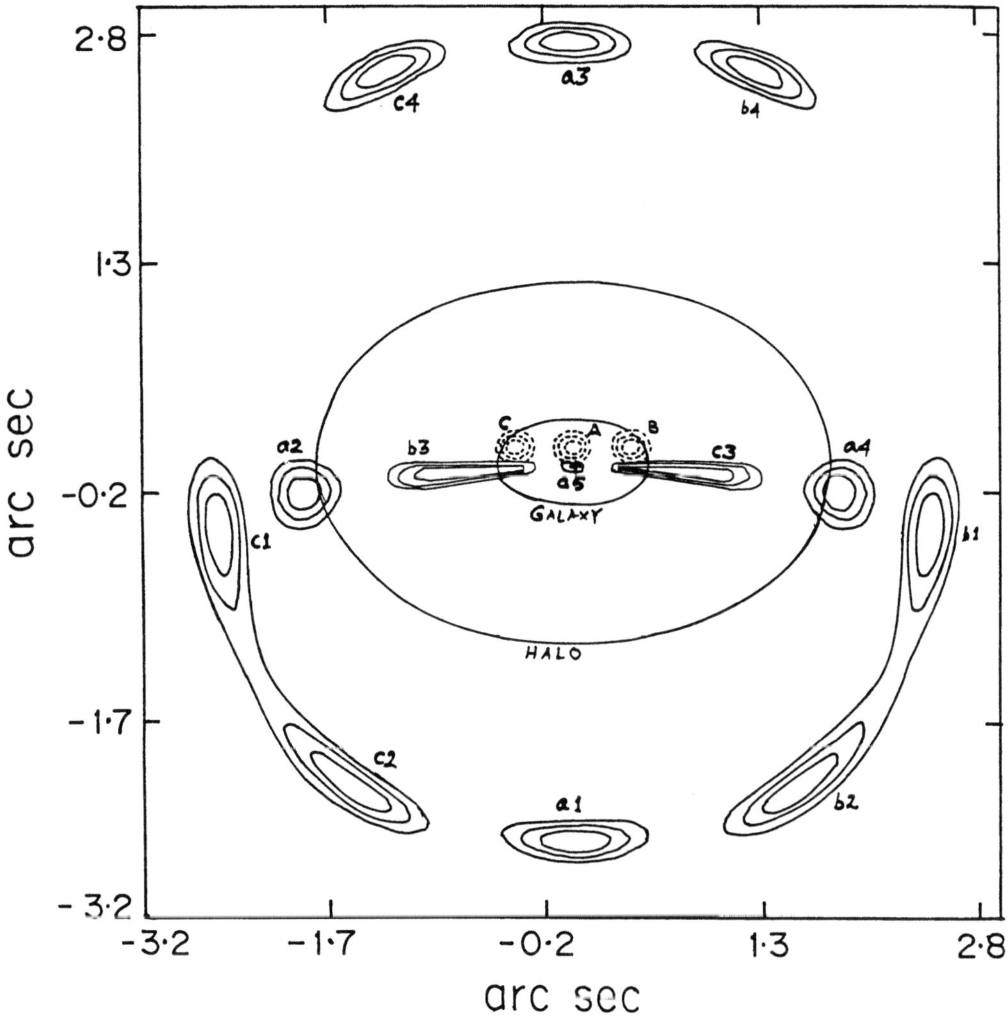

Fig. 2. The image configuration that can result when an extragalactic radio source made up of a central compact region (A) and two hot spots (B,C) located at a redshift of $z = 1$ is lensed by a foreground galaxy with a halo having a redshift of $z = 0.3$. The source contours are shown by dotted ellipses.

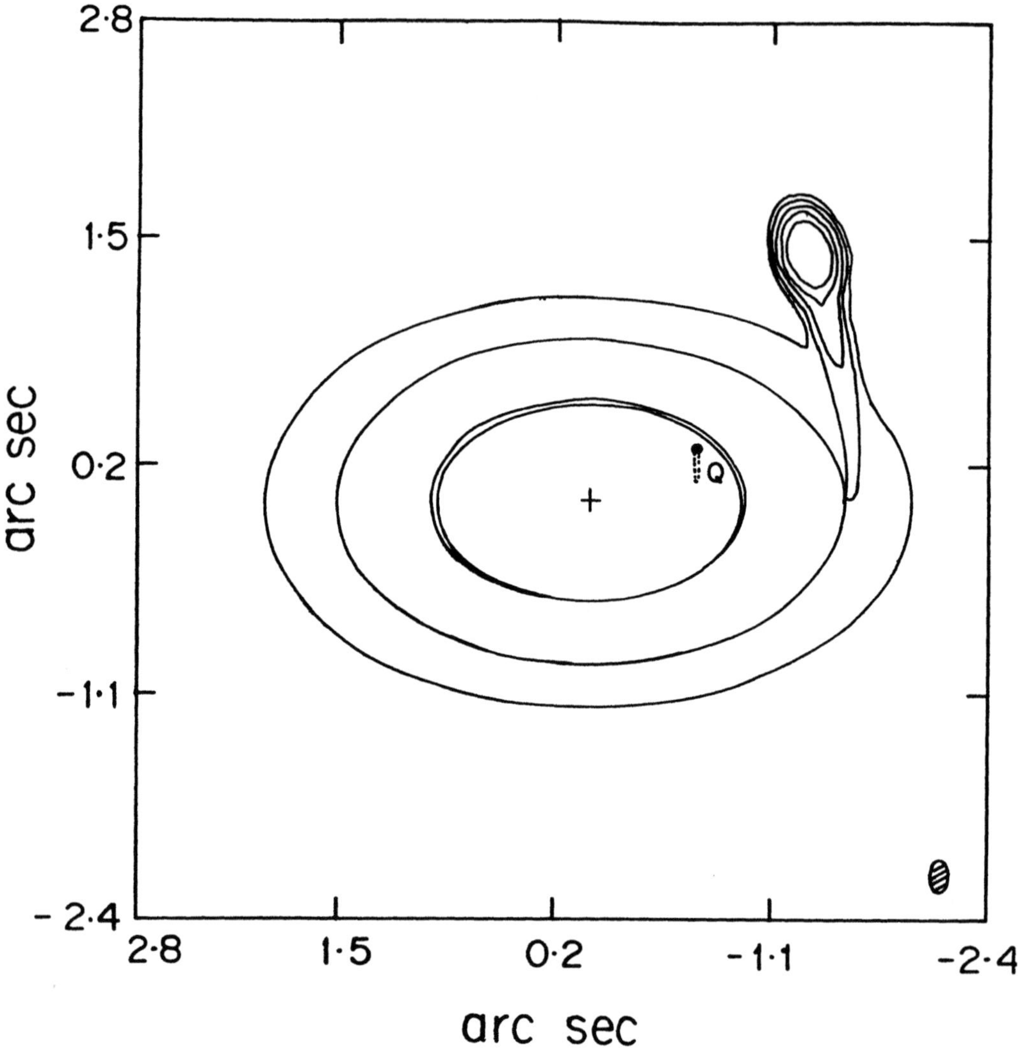

Fig. 3. Apparent luminous connection between a distant quasar (redshift $z = 1$) and a nearby galaxy (redshift $z = 0.04$) can be simulated if the source quasar with its jet indicated by a filled circle together with a dotted extension is lensed by the foreground galaxy shown with elliptical contours.

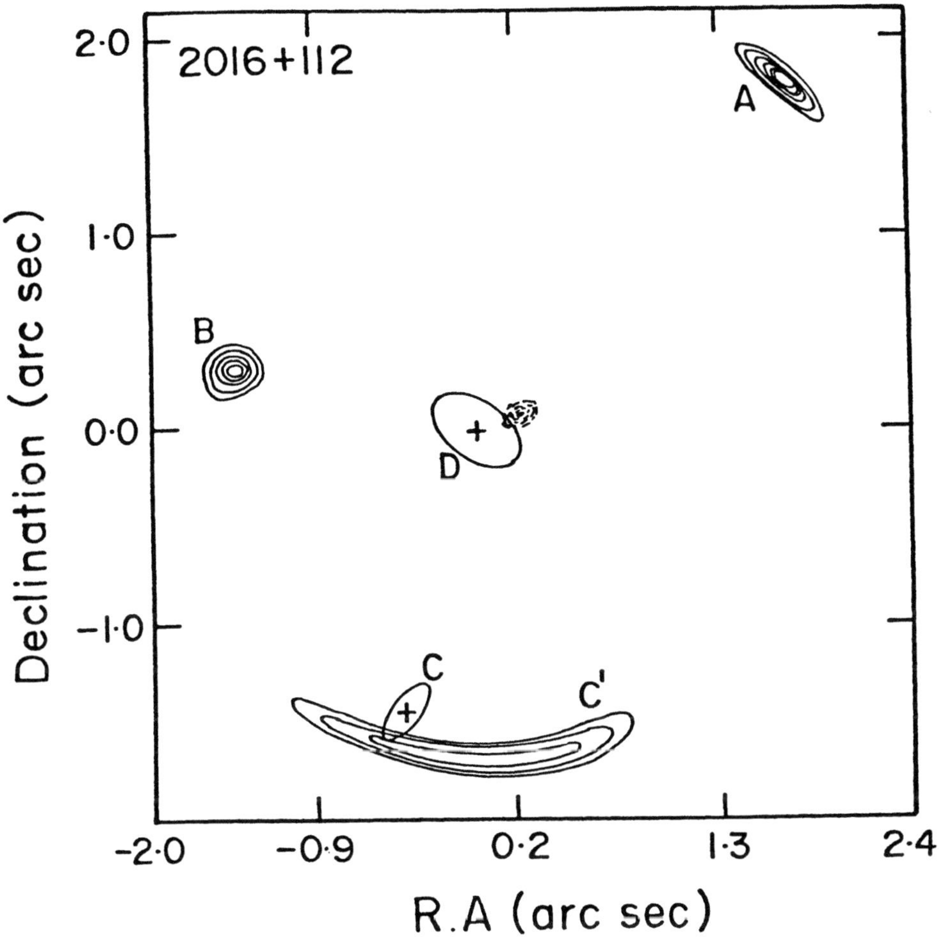

Fig. 4. The image configuration of 2016+112 is shown along with the lensing objects C and D indicated by solid ellipses with crosses at their center. The dotted curves denote contours of equal intensity of the background extended elliptical source.

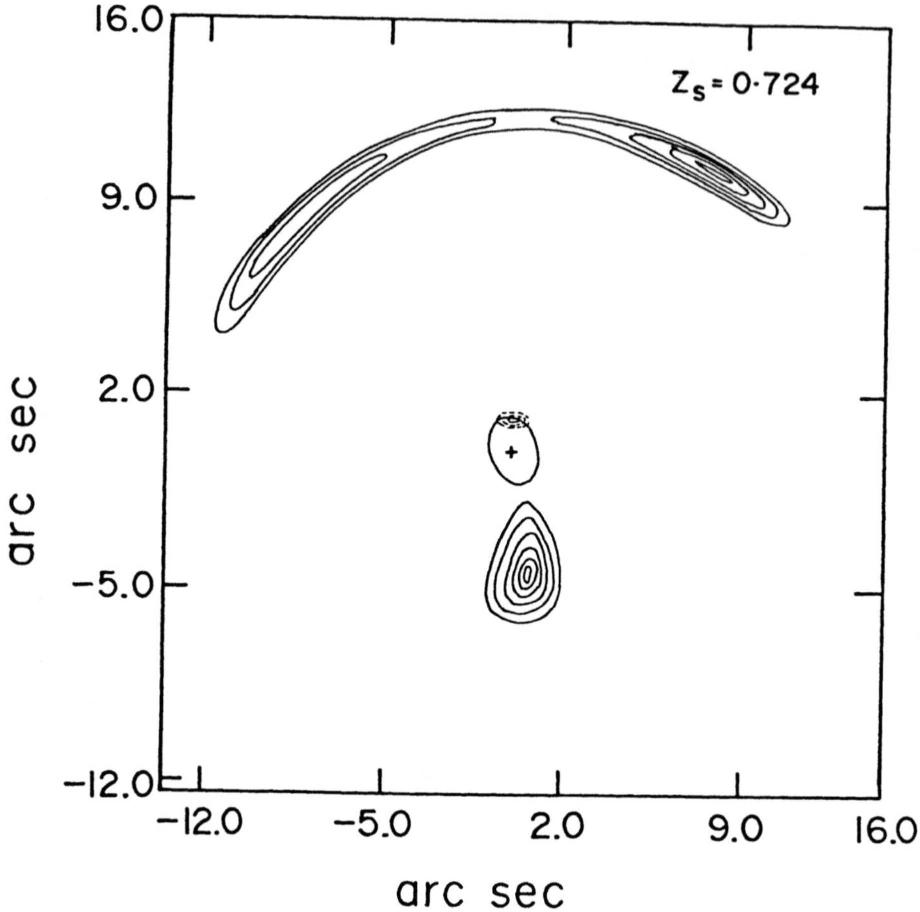

Fig. 5. Elliptical galaxy at a redshift $z_s = 0.724$ being lensed by a cluster of galaxies with a massive galaxy at its center with a redshift $z = 0.373$, to simulate the giant arc observed in the core of Abell 370. The lensing galaxy is shown by a solid ellipse with a cross at its center. The dotted curves indicate contours of equal intensity of the background source galaxy, while the solid elongated curves represent the corresponding image contours.

MOVING GRAVITATIONAL LENSES

M. Birkinshaw
Department of Astronomy, Harvard University
60 Garden Street, Cambridge, MA 02138, USA

Summary. A massive object distorts the path of light passing near it. If the object's metric changes at a different rate than that of the Universe, then a monochromatic source behind the object will change in frequency and flux relative to a similar source elsewhere. Contracting lenses and large masses in the early Universe provide effects of this type, but an alternative mechanism for producing such an effect is *motion* of the lensing object. A lensing object moving across the line of sight should cause (a) a redshift difference between multiple images of a background object (e.g., a quasar lensed by a galaxy), and (b) a brightness anisotropy in the microwave background radiation. Effect (a) is unlikely to be measurable for the multiple images formed by 'conventional' astrophysical objects, although a *string* may produce a detectable effect if it has sufficient mass per unit length. Effect (b) should soon be detectable for clusters of galaxies with large peculiar velocities.

Redshift Effects from Changing Gravitational Fields. Rees and Sciama (1968), Dyer (1976), Nottale (1984), and others have discussed the effect on background sources or the background radiation of a massive object surrounded by a void embedded in an expanding Universe. As light passes through the 'vacuole,' it is retarded relative to light that does not pass through the structure, so that a view of an anomalously early phase of the Universe is seen. If the vacuole expands or contracts at a different rate from the external Universe, light passing through it is redshifted differently from light that misses it. As a consequence, the brightness and spectrum of sources (and the background radiation) behind the vacuole are changed relative to other sources. This is a particular example of a more general phenomenon, in which optical effects are produced by localized changes in the metric that are unlike the changes in the metric of the Universe as a whole. A somewhat different effect arises when an object of fixed properties moves relative to the Hubble flow.

In the frame of a lens moving at velocity $\mathbf{v}$, the frequency of a monochromatic light source depends on the direction of the source relative to the direction of motion of the lens. The effect of the lens is to cause a change δ in the direction of a light ray without a change in frequency.

Back in the frame of the original source, a stationary observer would see that light emerging from the lens is no longer parallel to light entering the lens. The Lorentz transformations to and from the lens frame do not cancel out as they would have done had there been no deflection, and the observed frequency of the deflected light depends on δ and v, and the direction of the light source relative to $\mathbf{v}$. For nonrelativistic lens velocities and small angles of deflection, it is clear that the frequency shift should be of order $\delta \times v/c$. The exact result, for small deflections but arbitrary lens velocity, is

$$\frac{\Delta\nu}{\nu} = \gamma\beta\delta \sin\alpha \cos\phi \qquad (1)$$

where ν is the frequency of light emitted by the source, $\Delta\nu$ is the difference between the observed and emitted frequencies, $\beta = v/c$, $\gamma = (1-\beta^2)^{-1/2}$, α is the angle between the lens velocity vector ($\mathbf{v}$) and the line of sight to the lens, and ϕ is the angle between $\mathbf{v}$ and the emerging light ray projected onto the sky (Birkinshaw and Gull 1983). Note from the dependence on α that this effect arises from the *transverse* motion of the lens across the line of sight: measurement of $\Delta\nu$ provides a test for transverse motions of distant lensing objects.

Multiple Images of Quasars near Galaxies. Clearly the frequency shift given in equation (1) might be sought as a differential redshift between the multiple images of a quasar formed near an elliptical galaxy (Fig. 1). Consider the 0957+561 system as an example. In this case the separation of the A and B images is $\Delta\theta = 6.17$ arcsec (Gorenstein *et al.*1984), so that the frequency shift between lines in the spectra of A and B is $\frac{\Delta\nu}{\nu} \approx \frac{v}{c}\Delta\theta$, or a shift

$$\Delta\lambda \approx 5 \times 10^{-5}\,(v/1000\ \mathrm{km\,s^{-1}})\quad \mathrm{nm} \qquad (2)$$

in the wavelengths of optical lines. Since the narrowest line feature in a quasar spectrum has width $> 10^4\Delta\lambda$, it is clear that this shift is too small to be detected, even without considering

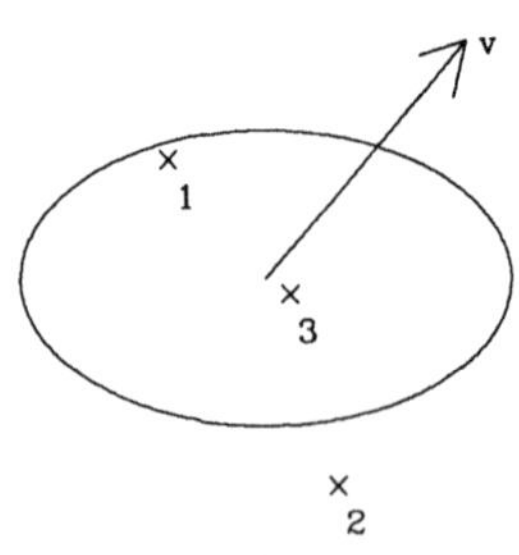

Fig. 1. A schematic diagram of the location of the three images (marked by numbered crosses) of a quasar formed by an elliptical galaxy (indicated by a representative isophote). The vector represents the direction of the peculiar velocity (v) of the galaxy projected in the plane of the sky.

the intrinsic differences that are likely because of the time delay and different lines of sight of the images.

The smallness of $\Delta\lambda$ can be understood easily as a consequence of the low density and velocity of most conventional astrophysical objects: if R_S is the Schwarzschild radius corresponding to the mass of the lens, the wavelength shift for a given image is

$$\frac{\Delta\lambda}{\lambda} \approx \frac{v}{c} \cdot \frac{R_S}{R} , \tag{3}$$

where R is the distance of closest approach of the ray to the center of the lens. Equation (3) makes it clear that $\Delta\lambda$ can become large only for rapidly moving, compact objects.

Strings and Multiple Images of Quasars. A much larger differential redshift should be seen in multiple images of quasars formed by cosmic strings (see Fig. 2). Suppose that the close pair of quasars 1146+111B and C represents an example of the lensing effect of a string (Turner *et al.*1986; Gott 1986). B and C are of similar brightness (as is required), and are separated by about 2.6 arcmin. This corresponds to a string with mass per unit length $\mu \approx 4 \times 10^{22}$ kg m^{-1} between the quasars, and the wavelength shift between the images is

$$\Delta\lambda \approx 0.4\,\gamma\beta \quad \text{nm} \tag{4}$$

for optical lines. Essentially all strings are expected to move at relativistic speeds, with $\beta\gamma \approx 1$. If this is true for a string in the 1146+111B,C system, a differential redshift $\Delta z \approx 0.001$ should

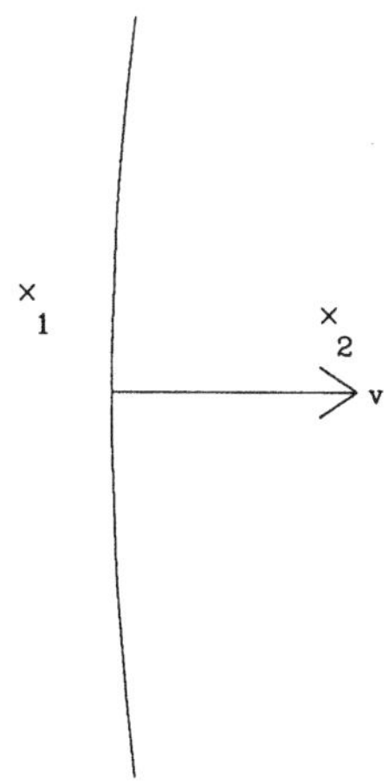

Fig. 2. A schematic diagram of the location of the two images (marked by numbered crosses) of a quasar formed by an string. Since the string is curved, it tends to move at relativistic speed (v), as indicated by the vector.

appear between the B and C images. A redshift difference of this size might also appear if the quasars are independent members of a single cluster. (Note that a second contribution to a redshift difference arises from the different time delays for the two images, but this is of order δ^2, and hence is smaller than the velocity-produced delay above; Gott 1985).

If the Universe holds strings with μ greater than the value suggested for 1146+111B,C, the redshifts of the two images of a lensed quasar might be distinctly different, and the lensing system might be difficult to recognize. The similarity in the quasar spectra and brightnesses may be further reduced by differences in the absorbing medium along the line of sight, changes in the quasar spectrum over the differential light travel time, etc., so that there may be significant problems in using optical search methods to locate strings.

Background Radiation Anisotropy. Figure 3 illustrates the theoretical pattern of frequency shifts that is expected around a moving lens with a density distribution of the form

$$\rho(\mathbf{r}) = \begin{cases} \left(1 + r^2/r_c^2\right)^{-3/2} & r < r_0 \\ 0 & r > r_0 \end{cases}$$

with $r_0 = 4r_c$. As indicated earlier, the frequency shifts themselves will probably not be directly observable in the redshifts of multiply imaged quasars, but the associated changes in the brightness of a background radiation may be observable: light from a uniform background is preferentially shifted to higher frequencies on one side of the lens, lower frequencies on the other, so that a brightness change is seen across the lens. When account is taken of the change in solid

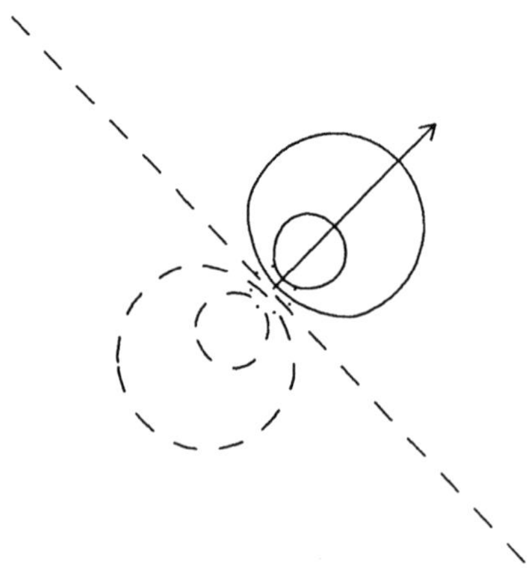

Fig. 3. The frequency shift $\Delta\nu/\nu$ produced by a simple mass distribution. The contours are drawn at -1.0, -0.5, 0.0, 0.5 and 1.0 in units of $\frac{4GM}{r_0 c^2}\gamma\beta\sin\alpha$, and negative and zero contours are drawn dashed. The small dotted circle represents the core radius of the mass distribution, and the vector indicates the direction of motion.

angle occupied by a bundle of rays (Mitrofanov 1981) as well as the frequency shift, it can be shown that the brightness changes by a fraction

$$\frac{\Delta B}{B} = -\frac{\Delta \nu}{\nu} \quad . \tag{5}$$

[Note that Birkinshaw and Gull (1983) neglected the transformation of solid angle and hence obtained a result in error by a factor -2.]

The deflection angle δ is relatively large for a rich cluster of galaxies, about 1 arcmin. The transverse velocities of clusters of galaxies are unknown, but might be as large as 1000 $\mathrm{km\,s^{-1}}$ (Dressler *et al.* 1987). Then the peak-to-peak change in the brightness temperature of the microwave background radiation across a moving cluster of galaxies is

$$\Delta \mathrm{T} \approx 10\,(v/1000\ \mathrm{km\,s^{-1}}) \quad \mu\mathrm{K}. \tag{6}$$

Present observations of the microwave background radiation with single dishes achieve rms errors $\approx$ 30 μK, so that the signals from moving clusters should be measurable in the near future. The detectability of this effect is enhanced, however, by its unusual angular structure (Fig. 3), and interferometric strategies may be effective for this reason.

Other moving masses, for example, galaxies or cosmic strings, will produce a similar type of signal in the microwave background radiation. For a galaxy, the signal will be smaller by a factor $\sim 10^2$ in brightness and will appear on an angular scale of a few seconds of arc. Although the angular scale is then a good match to the capabilities of interferometers such as the VLA, the sensitivity required to detect this signal is not available at present.

A moving string should produce a large step in the microwave background radiation (Kaiser and Stebbins 1984): for 1146+111B,C the brightness temperature change is $\Delta\mathrm{T} \approx 2\beta$ mK. Such signals should be detected easily using present single-dish techniques if $\beta > 0.1$. The absence of any such brightness signature in the region near the quasars 1146+111B and C was therefore interpreted as strong evidence against the interpretation of this system in terms of a cosmic string (Lawrence *et al.* 1986; Stark *et al.* 1986).

References.

Birkinshaw, M. and Gull, S.F., 1983. *Nature*, **302**, 315.

Dressler, A., Faber, S.M., Burstein, D., Davies, R.L., Lynden-Bell, D., Terlevich, R.J. and Wegner, G., 1987. *Ap. J.*, **313**, L37.

Dyer, C.C., 1976. *M. N. R. A. S.*, **175**, 429.

Gorenstein, M.V,. Shapiro, I.I., Rogers, A.E.E., Cohen, N.L., Corey, B.E., Porcas, R.W., Falco, E.E., Bonometti, R.J., Preston, R.A., Rius, A. and Whitney, A.R., 1984. *Ap. J.*, **287**, 538.

Gott, J.R., 1985. *Ap. J.*, **288**, 422.

Gott, J.R., 1986. *Nature*, **321**, 420.

Kaiser, N. and Stebbins, A., 1984. *Nature*, **310**, 391.

Lawrence, C.R., Readhead, A.C.S., Moffet, A.T. and Birkinshaw, M., 1986. *A. J.*, **92**, 1235.

Mitrofanov, I.G., 1981. *Pis'ma Astr. Zh.*, **7**, 73.

Nottale, L., 1984. *M. N. R. A. S.*, **206**, 713.

Rees, M.J. and Sciama, D.W., 1968. *Nature*, **217**, 511.

Stark, A.A., Dragovan, M., Wilson, R.W. and Gott, J.R., 1986. *Nature*, **322**, 805.

Turner, E.L., Schneider, D.P., Burke, B.F., Hewitt, J.N., Langston, G.I., Gunn, J.E., Lawrence, C.R. and Schmidt, M., 1986. *Nature*, **321**, 142.

EXPLAINING BURKE AND SHAPIRO TO NEWTON

David H. Frisch
Physics Department
Massachusetts Institute of Technology
Cambridge, MA 02139

Newton kept asking about an "aetherial medium" in or near "material bodies" that could account for the refraction of light. We now know that the strong refractive effects that Newton measured come from the reradiation of light by electrons in the material bodies. In addition, there is a very weak effect of matter that can be treated as a perturbation of Newtonian space and time themselves: For every space interval ds, substitute

$$ds' \cong ds\left(1 - \frac{GM}{c^2 r}\right) ,$$

and for a time interval dt, substitute

$$dt' \cong dt\left(1 + \frac{GM}{c^2 r}\right) .$$

These give a speed field, for everything,

$$\frac{v'}{v}\left(= \frac{c'}{c}!\right) \cong 1 - \frac{2GM}{c^2 r} ,$$

which allows near-trivial perturbation calculation of the bending of light, the delay of a radar echo, and even the entrainment of an orbiting gyro's spin direction by rotation of the earth. On adding a special relativistic correction (Thomas Precession), the De Sitter gyro precession is directly calculable; and special relativity (Sommerfeld Precession) plus gravitation by energy $-GMm/r$ in the field account for the precession of orbital perihelia.

IV. Observations of Gravitational Lenses

RECENT OPTICAL OBSERVATIONS OF GRAVITATIONAL LENSES

Edwin L. Turner
Princeton University Observatory
Peyton Hall, Ivy Lane, Princeton, NJ 08544-1001

Introduction. Due, in significant part, to the stimulating effect of Bernie Burke's energy and leadership during recent years, there is no hope of my presenting even a marginally complete survey of important recent optical observations of gravitational lenses in the available time and space, even though I can grudgingly leave aside the exciting giant luminous arcs, which are treated separately elsewhere in this volume. I shall not try. Instead, this paper will concentrate on a few recent results that bear on specific scientific issues that happen to have particularly attracted my interest. These issues include 1) the problem of obtaining unique or at least well-constrained models of lens mass distributions, 2) the task of distinguishing real gravitational lens systems from physical pairs and multiples, 3) information on dark matter and its distribution, and 4) the discovery of new lens systems. In these connections, I will discuss recent optical observations of 2237+0305, 1635+267, 2345+007, and MG0414+0534 as well as arecent optical imaging survey of a sample of high luminosity quasars.

2237+0305. This is the lens system discovered serendipitously in the course of the CfA Redshift Survey by Huchra *et al.* (1985) in which a redshift 1.69 quasar is multiply imaged by the nucleus of a redshift 0.04 barred spiral galaxy. Extensive image analysis and decomposition of high signal-to-noise CCD frames of the system obtained at Mauna Kea and KPNO by Yee (1988) and Schneider *et al.* (1988), respectively, have yielded a detailed quantitative description of this lens systems configuration. This includes the relative positions, apparent magnitudes, and colors of four images of the quasar and of the galaxy's nucleus as well as a measure of the galaxy's surface brightness distribution. The excellent agreement of the results of the Yee (1988) and the Schneider *et al.* (1988) studies, which were carried out independently, based on different data sets, and used somewhat different iterative image decomposition techniques, gives considerable confidence in their validity.

Schneider *et al.* (1988) used these data to construct a detailed numerical gravitational lens model of 2237+0305, which is extremely well constrained and may be reasonably considered unique. Their procedure was to assume that the observed surface brightness distribution of the galaxy is directly proportional to its surface mass distribution, i.e., a constant mass-to-light ratio (probably reasonable for the nuclear regions of a galaxy). The task was then to explain the 12 known observables (x, y position and brightness for each of the four images) with the four remaining lens model parameters (unperturbed x, y position and brightness of the quasar and the galaxy M/L value). It turned out to be possible to account for all 12 of the observables within their measurement errors with a particular, well-determined set of the four parameters. Since this success was by no means guaranteed (or even likely) if the true mass distribution did not pretty closely match the observed galaxy surface brightness distribution, one can reasonably have fair confidence in the validity of the Schneider *et al.* (1988) model.

This result is important for several reasons: First, it validates the gravitational lens interpretation and the conventional cosmological explanation of redshifts, which have been questioned (Burbidge 1985) for this striking system. Second, it gives a novel and rather precise measure of the central mass and mass-to-light ratio of the galaxy, which are in general agreement with those determined by conventional dynamical studies of many other galaxies; it thus amounts to a verification of the traditional dynamical analyses. Third, it raises the possibility of using the system for determination of cosmological parameters such as Hubble's constant without undue model dependency. Fourth and finally, it provides an ideal system in which to search for and study microlensing events: the indicated microlensing "optical depths" are in the optimum range of several tenths, the mass along the four lines of sight through the galaxy is almost certainly dominated by objects of stellar mass, and the surprisingly low redshift of the lensing galaxy implies a minimally restrictive limit on the true angular size of the quasar image for effective microlensing.

<u>1635+267.</u> This system is an excellent example of an ambiguous gravitational lens candidate; it consists of two quasar images showing reasonably similar emission line spectra, clear differences in continuum shape, and no measurable velocity difference. Attempts to locate a foreground lensing galaxy have yielded only fairly low limits on its brightness. Although interpretation of 1635+267 as a physical pair was considered "very unlikely" by its discoverers (Djorgovski and Spinrad 1984), other authors (Blandford and Kochanek 1988) have reached the opposite conclusion on the basis of the same data considering it "probably" to be a physical pair. The task of distinguishing between real lens systems and physical pairs (or multiples) is probably the most pressing and difficult general observational problem now facing gravitational lens studies. Although the cases with detected lensing objects are relatively clear cut, it is very undesirable to automatically exclude the possibility of detecting lensing by dark objects, potentially one of the most important classes of lenses.

In order to address this problem for 1635+267 and perhaps to develop a more generally useful technique, Turner *et al.* (1988) obtained high signal-to-noise spectra of its two components and, after interpolating removal of the continua in each spectrum, compared the emission lines of Mg II and C III] in each image. Without recounting the details of the published study, their basic result can be stated as the determination that, to within measuring errors, the wavelengths, strengths, widths, and profile shapes (including detectable asymmetries) of each line match those of that line in the other image aside from a single overall scaling factor. This factor can be reasonably interpreted as a magnification ratio in the lensing event. Moreover, it was shown that such a good match is not present in a (single) unrelated quasar of similar redshift and apparent magnitude. Turner *et al.* (1988) also detect possible evidence for redshift 0.57 galaxy absorption features in the continuum spectrum of the brighter and redder image. These spectroscopic results lend further support to the gravitational lens hypothesis for 1635+267 (although they cannot be considered entirely decisive) and suggest a technique of high signal-to-noise spectroscopy that may be useful for other uncertain gravitational lens candidates.

2345+007. Since its discovery (Weedman *et al.* 1982), 2345+007 has been a particularly intriguing but controversial gravitational lens candidate. Despite its large image separation of 7″, very deep CCD images in excellent seeing have failed to reveal evidence of a lensing galaxy or cluster of galaxies (Tyson *et al.* 1986). Again, opinions have differed over whether the evidence favors the lens hypothesis or the physical pair interpretation.

Recently, Nieto *et al.* (1988) have obtained very high resolution CCD images of the system under conditions of excellent seeing (FWHM = 0.4″). By using addition and recentering of individual exposures plus image restoration (or super resolution) techniques, they achieved good signal-to-noise images with FWHM of roughly 0.″25. These superb images show that the fainter (B component) of the previously known two images is, in fact, a double separated by about 0.″4 roughly along the same direction as the primary A and B images. Furthermore, there is some evidence that the outer of these two subimages is resolved in an azimuthal direction.

Few, if any, ordinary quasars have been subjected to such high resolution imaging so that it is impossible to say how common such image structure is in unlensed quasars. Nevertheless, since the gravitational lens hypothesis for 2345+007 generically predicts additional source images and azimuthal distortions, it is logical to conclude that the new data support the lensing hypothesis on their face and require more complex (and perhaps contrived) scenarios in the multiple object explanation.

MG0414+0534. Elsewhere in this volume, Hewitt *et al.* (1989) report on a new and very promising gravitational lens candidate, MG0414+0534, discovered in the ongoing VLA gravitational lens survey. Images of this system in the radio U and C bands and in the optical/near-IR g, R, I, and Z bands all show exactly the same structure. Moreover, the image configuration is a standard one expected for lensing of a point source by a foreground elliptical potential [see Narayan and Grossman (1989) elsewhere in this volume, for example].

The spectra of the components of this object considered together with the deep, good-seeing CCD images are quite perplexing. The spectra show an extraordinarily red continuum between about 5000 and 9000 angstroms with no detectable emission or absorption features aside from a possible break that could be interpreted as the Ca II H and K break at a redshift of approximately 1.2. The CCD images, which have good resolution (better than 1″) and extend well into the red and near-IR, show no hint of extended emission such as would be expected from a galaxy (either the lens or associated with the source) at any likely redshift. Furthermore, the point components in the source are quite faint (r of about 23) and unlikely to be able to swamp any extended emission.

Thus, although the evidence supporting the lensing hypothesis for MG0414+0534 is reasonably convincing, neither the source nor the lens appears to be a member of any familiar class of astronomical objects. Clearly, this system warrants more attention.

Luminous Quasar Survey. Surdej *et al.* (1988) have recently reported an imaging survey of 111 quasars with calculated absolute visual magnitudes brighter than –29, the idea being that their luminosities might be due in part to gravitational lens magnification. Twenty-five of these quasars showed nonstellar images ranging from slight "fuzz" through image elongation to detected multiple images. Five of the twenty-five were considered by the investigators to be "excellent" gravitational lens candidates, and two of these have already been confirmed spectroscopicly (Surdej *et al.* 1987, Magain *et al.* 1988). The most striking statistical result of this survey is the strong correlation of detected image structure with the seeing conditions. In particular, four of the five "excellent" lens candidates were found among the 40 images with the best seeing and three were among the 11 images with seeing FWHM better than 0.″9. This result strongly suggests that a much larger fraction of the luminous quasar sample would show possible evidence of lensing if they could all be imaged with high resolution. Certainly one's appetite for Hubble Space Telescope images of luminous quasars is greatly whetted. If the implications of the Surdej *et al.* (1988) result are borne out, they not only promise a new, high efficiency technique for locating lens systems but also suggest that lensing magnifications may strongly affect quasar population statistics, at least in the high luminosity regime.

Discussion. Although this review has been oriented only toward optical observations, has concentrated almost exclusively on very recent results (mostly less than six months old), and has omitted one of the most important new topics (luminous arcs), it is still quite incomplete. This situation certainly attests to the intense current level of activity in gravitational lens studies. But, what have the new results described above taught us about the four scientific issues of particular interest enumerated in the first section?

First, it does appear that unique, well-constrained models of the lens mass distributions can be obtained in at least some favorable situations, as demonstrated by 2237+0305.

Second, although the problem of distinguishing real gravitational lens systems from physical pairs and multiples is likely to remain a thorny and controversial issue, recent observational studies of systems (1635+267 and 2345+07) that have been considered by some to be among the most dubious candidates have generally lent new support to the gravitational lens interpretations.

Third, on the topic of dark matter, gravitational lenses continue primarily to give us some reassurances about conventional views: they verify standard accounts of the mass distributions in the centers of galaxies (2237+0305); the fact that most lens candidates do have detectable lensing galaxies implies that the Universe does not contain a population of dark compact objects of roughly galaxian mass that are more abundant than galaxies, even though there remain a few intriguing candidates for dark lenses (1635+267, 2345+007, and perhaps MG0414+0534).

Fourth, new and intriguing lens systems continue to be discovered using established techniques (MG0414+0534), and new and potentially extremely important search techniques are being developed (high luminosity quasar imaging).

Concluding on a note that is, I think, characteristic of Bernie Burke's approach to science (and life), developments in all four of the scientific areas discussed here should give us cause for great optimism and motivation to press on with the jobs at hand.

Acknowledgments. This work was supported in part by NSF grant AST86-18257. The spectroscopic and imaging data on MG0414+0534 discussed here were obtained in collaboration with B. F. Burke, J. E. Gunn, J. N. Hewitt, C. R. Lawrence, D. P. Schneider, and M. Schmidt, but they cannot be held responsible in any way for my interpretive comments. I would like to thank Bernie Burke who did not introduce me to gravitational lens studies but who drew me, in that irresistible way so familiar to all his friends, into concentrating more and more of my efforts on them; I have enjoyed it a great deal.

References.

Blandford, R. D., and Kochanek, C. S. 1988, preprint.

Burbidge, G. 1985, *A. J.*, **90**, 1399.

Djorgovski, S., and Spinrad, H. 1984, *Ap. J. (Letters)*, **282**, L1.

Hewitt, J. N., Burke, B. F., Turner, E. L., Schneider, D. P., Lawrence, C. R., Langston, G. I., and Brody, J. P. 1989, this volume.

Huchra, J., Gorenstein, M., Kent, S., Shapiro, I., Smith, G., Horine, E., and Perley, R. 1985, *A. J.*, **90**, 691.

Magain, P., Surdej, J., Swings, J. P., Borgeest, U., Kayser, R., Kuhr, H., Refsdal, S., and Remy, M. 1988, *Nature*, **334**, 325.

Narayan, R. and Grossman, S. 1989, this volume.

Nieto, J. L., Roques, S., Llebaria, A., Vanderriest, Ch., Lelievre, G., di Serego Alighieri, S., Macchetto, F. D., and Perryman, M. A. C. 1988, *Ap. J.*, **325**, 644.

Schneider, D. P., Turner, E. L., Gunn, J. E., Hewitt, J. N., Schmidt, M., and Lawrence, C. R. 1988, *A. J.*, **95**, 1619.

Surdej, J., Magain, P., Swings, J. P., Borgeest, U., Courvoisier, T. J. L., Kayser, R., Kellermann, K. I., Kuhr, H., and Refsdal, S. 1987, *Nature*, **329**, 695.

Surdej, J., Swings, J. P., Magain, P., Borgeest, U., Kayser, R., Refsdal, S., Courvoisier, T. J. L., Kellermann, K. I., and Kuhr, H. 1988, in *Proceedings of a Workshop on Optical Surveys for Quasars*, ed. P. S. Osmer, A. C. Porter, R. F. Green, and C. B. Foltz (Astronomical Society of the Pacific, San Francisco), p. 183.

Turner, E. L., Hillenbrand, L. A., Schneider, D. P., Hewitt, J. N., and Burke, B. F. 1988, *A. J.*, **96**, 1682.

Tyson, J. A., Seitzer, P., Weymann, R. J., and Folz, C. 1986, *A. J.*, **91**, 1274.

Weedman, D. W., Weymann, R. J., Green, R. F., and Heckman, T. M. 1982, *Ap. J. (Letters)*, **255**, L5.

Yee, H. K. C. 1988, *A. J.*, **95**, 1331.

VLBI OBSERVATIONS OF GRAVITATIONAL LENSES

Marc V. Gorenstein[1]
Harvard–Smithsonian Center for Astrophysics
60 Garden Street, Cambridge, MA 02138 USA

Current Observations. To date, six publications have appeared in refereed journals that describe observations of gravitational lens systems using the technique of Very-Long-Baseline Interferometry (VLBI): Gorenstein *et al.* (1983, 1984, 1988a); Haschick *et al.* (1981); and Porcas *et al.* (1979, 1981). All are on observations of the images of 0957+561, the first identified lens system.

The primary contribution of these VLBI observations to the study of gravitational lenses is the demonstration that the brightness distributions of the quasars 0957+561A and B are consistent with their being images of the same object. In fact, Gorenstein *et al.* (1988a) obtained the four components of the relative magnification matrix that relate the brightness distributions of the large scale features of the A image to those of the B image.

VLBI also has the potential to determine the angular separation between images and the lens (if the latter is radio loud), to detect the existence of other images, and to monitor the evolution of the brightness distribution of images (Gorenstein *et al.* 1988b).

Heflin *et al.* (1988) described preliminary results from VLBI observations of 2016+112. The results are consistent with the lens hypothesis; however, because of the poor *u-v* coverage, due to the system's low declination, the uniqueness of the brightness models is not certain.

Future Prospects. There are only some dozen known or suspected lens systems, and of these only the images of 0957+561 and 2016+112 have sufficient flux density (around 50 mJy per image at 18 cm wavelength) to be mapped or modeled with VLBI.

More examples of radio loud lensed quasars are needed as the utility of such observations seems clear: in the case of 0957+561, the values of the components of the magnification matrix

[1] Current address: Waters Chromatography Division, Millipore Corporation, 34 Maple Street, Milford, MA 01757

can be used to constrain mass models of the lenses, and measurements of the time evolution of the radio brightness distributions of images can lead to a measurement of the Hubble constant, H_0 (see, e.g., Falco, Gorenstein, and Shapiro 1988a,b).

The large systematic VLA-based radio surveys pioneered by B. F. Burke have discovered radio loud lenses [see Hewitt, this volume, and Lawrence *et al.* (1984) for the case of 2016+112] and promise to provide new candidates for VLBI observations. The VLB Array, now under construction, will make the acquisition and analysis of VLBI data of weak sources significantly more convenient. We can anticipate that the VLBI technique can make significant new contributions to the study of lens systems.

References.

Falco, E. E., Gorenstein, M. V., and Shapiro, I. I. 1988a, in *The Impact of VLBI on Astrophysics and Geophysics, Proceedings of IAU Symposium No. 129*, ed. M. J. Reid and J. M. Moran (Dordrecht: Reidel), p. 207.

Falco, E. E., Gorenstein, M. V., and Shapiro, I. I. 1988b, in preparation.

Gorenstein, M. V., *et al.* 1983, *Science*, **219**, 54.

Gorenstein, M. V., *et al.* 1984, *Ap. J.*, **287**, 538.

Gorenstein, M. V., Cohen, N. L., Shapiro, I. I., Rogers, A. E. E., Bonometti, R. J., Falco, E. E., Bartel, N., and Marcaide, J. M. 1988a, *Ap. J.*, **334**, 42.

Gorenstein M. V., *et al.* 1988b, in *The Impact of VLBI on Astrophysics and Geophysics, Proceedings of IAU Symposium No. 129*, ed. M. J. Reid and J. M. Moran (Dordrecht: Reidel), p. 201.

Haschick, A. D., Moran, J. M., Reid, M. J., Davis, M. J., and Lilley, A. E. 1981, *Ap. J. (Letters)*, **243**, L57.

Heflin, M., *et al.* 1988, in *The Impact of VLBI on Astrophysics and Geophysics, Proceedings of IAU Symposium No. 129*, ed. M. J. Reid and J. M. Moran (Dordrecht: Reidel), p. 209.

Lawrence, C. R., *et al.* 1984, *Science*, **223**, 46.

Porcas, R. W., Booth, R. S., Browne, I. W. A., Walsh, D., and Wilkinson, P. N. 1979, *Nature*, **282**, 385.

Porcas, R. W., Booth, R. S., Browne, I. W. A., Walsh, D., and Wilkinson, P. N. 1981, *Nature*, **289**, 758.

VLBI PHASE REFERENCE MAPPING TECHNIQUES AND THE SEARCH FOR THE THIRD IMAGE OF 0957+561

Alan E. E. Rogers
Haystack Observatory
Westford, MA 01886

Abstract. This paper reviews the technique of using a phase reference in VLBI to extend the coherence and to provide accurate differential astrometry and describes the search for the third image of 0957+561. The search detected a compact component that was initially seen at 2.3 GHz and reported by Gorenstein *et al.* (1983). Later this compact source was also observed at 8.4 GHz and 5 GHz. Whether this source is the long sought third image or the core of the lensing galaxy has been an open question for several years. The interferometer phases on the most sensitive VLBI baseline suggest that we have observed the third image.

Phase Reference Mapping Techniques. In simple terms, phase reference mapping uses a strong calibration source to provide a phase reference so that phase fluctuations from the clock and atmosphere are cancelled, thereby making the interferometer phase stable. For a perfect phase reference, a correction for the structure phase of the reference source has to be made. In the case of the gravitational lens system 0957+561, the reference source was chosen to be the "B" image, which, being about an arcsecond from the search region, was observed simultaneously. In this way, the region was "self" referenced and therefore did not require any interpolation or extrapolation of the reference phase as is required when the reference source is not so close and has to be observed by switching the antenna beams back and forth between signal and reference regions of the sky. In principle, the mapping or search can be done by the conventional method of Fourier transformation of the complex phase referenced data. However, the very large values of u and v, combined with the small separation of the uv points sampled every two seconds in time and every 2 MHz in frequency, requires the use of very large transforms and storage arrays. A much simpler method was used in which the complex phase referenced data for each scan was transformed (using discrete "zoom" transform) to the sky plane and then averaged in the sky plane. After having detected a significant peak in the averaged map from all scans (and baselines), the amplitude and phase for each scan offset to the position of the peak can be mapped and cleaned in the conventional manner. The complex amplitude for each scan is an average in time and frequency, or equivalently in u and v, as illustrated in Figure 1 and is thereby filtered by the delay/rate beam. A wider field could be mapped by increasing the size of

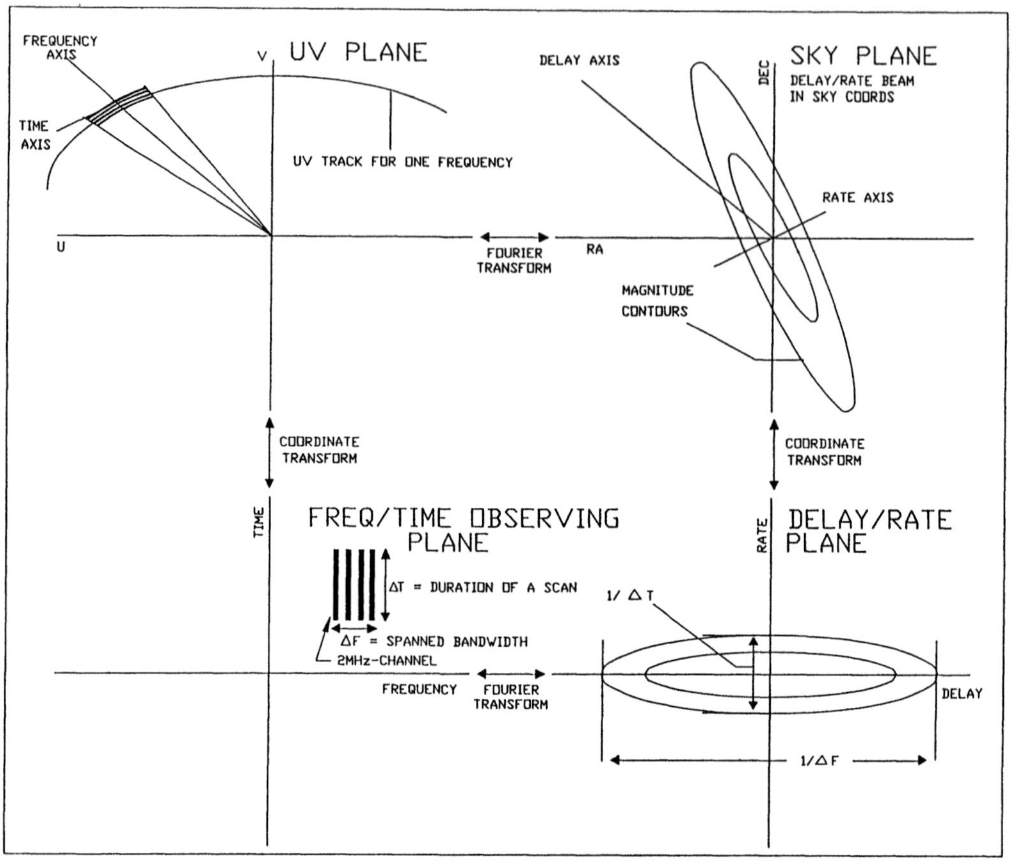

Fig. 1. The relationship between the *uv*, frequency/time, delay/rate and sky planes. The 0957+561 experiments used Mark III VLBI system to record 28 2-MHz channels.

the delay/rate beam. This can be accomplished by reducing the scan length and/or segmenting into subsets of frequencies.

The Compact Source in 0957+561. We initially searched a region 0.2 arc sec North, 0.2 arc sec East, 0.2 arc sec West, and 0.6 arc sec South of the VLA position for the Galaxy G. The only significant signal was at a position 0.181 arc sec East and 1.029 arc sec North of the VLBI core of the B image. Figure 2 shows a schematic illustration of the 0957+561 region and the cleaned map of the compact VLBI component obtained from March 1981 data on the Bonn-Madrid-Goldstone triangle. While there is a suggestion of a jet, the SNR in this component is only 3 as compared with an overall SNR of about 15 in the compact component. With exception of the mention of the possible jet, the results were published by Gorenstein *et al.* (1983). 0957+561 was again observed in May 1983 with the same antennas at both 2.3 and 8.4 GHz, and the core of the compact component had increased by a factor of 2 at 2.3 GHz and was detected

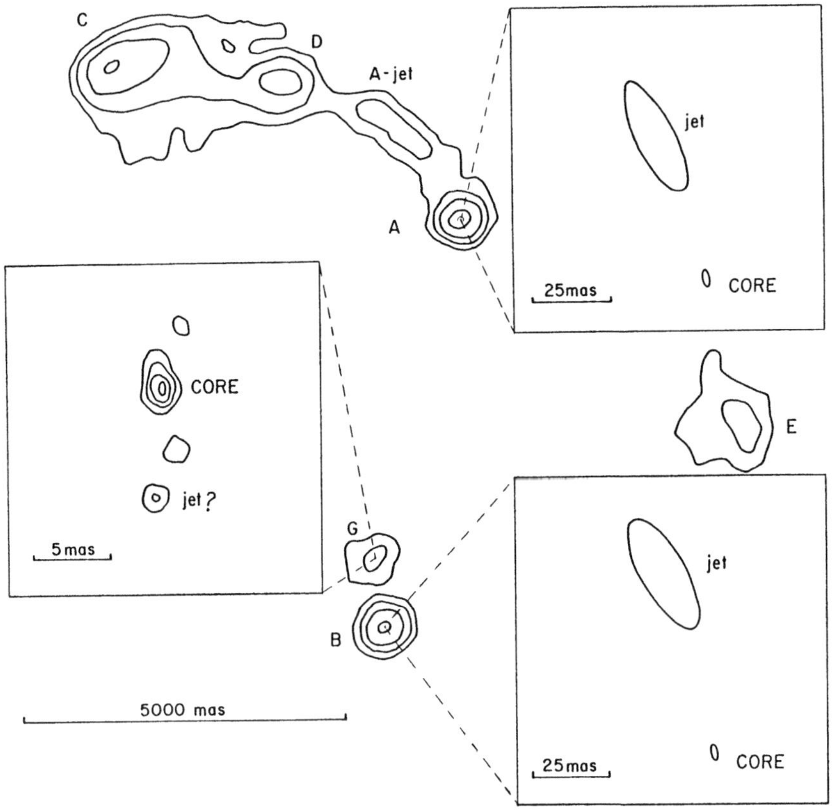

Fig. 2. A schematic of the 0957+561 lens system with map of the compact VLBI source (which may be the third image) made from 1981 data on the Bonn-Madrid-Goldstone triangle – phase referenced to the B component.

at 8.4 GHz. Table 1 summarizes the strength of this compact component which we call G'. Another experiment in July 1983 at 5 GHz between Bonn and the VLA also detected the compact components (see Bonometti 1985) with a strength of about B/30.

Table 1
Strength of G′ on the Goldstone–Madrid Baseline
(ΔR.A., ΔDec. = 181,1029 mas)

Date of Observation	G' at 13 cm	G' at 3.8 cm
March 1981	B/36	–
May 1983	B/18	B/10

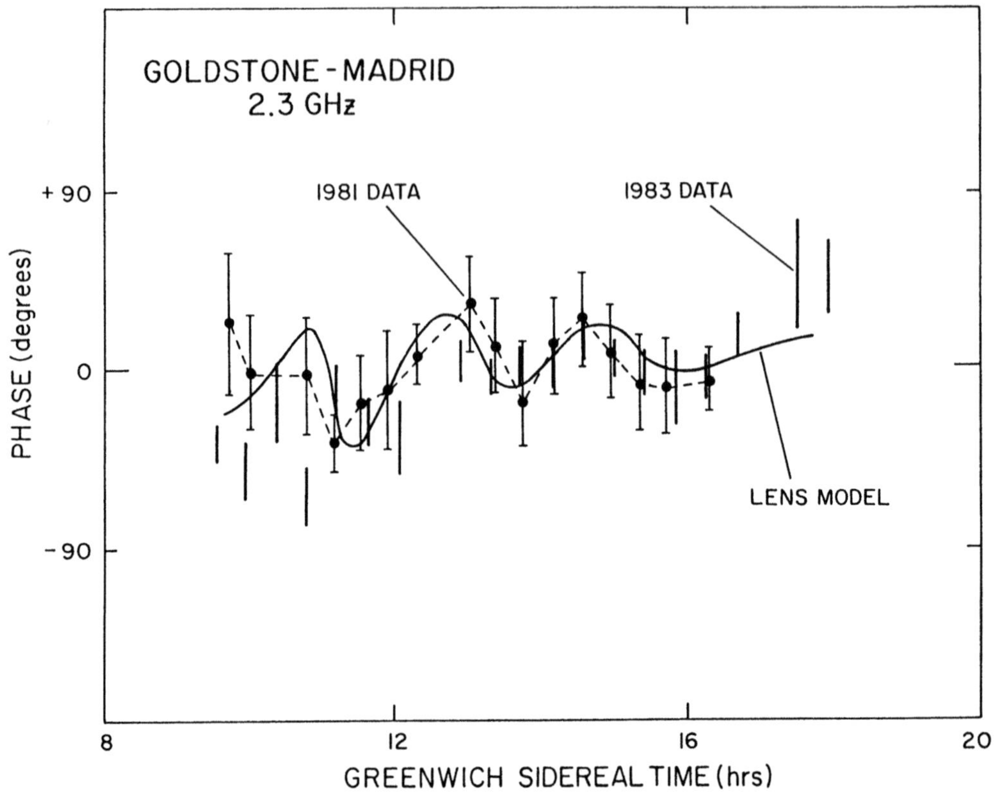

Fig. 3. The phase of G' observed on the most sensitive Goldstone-Madrid baseline at 2.3 GHz. The lens model (component A demagnified by a factor of 6) and 1983 data are taken from Robert Bonometti's 1985 Ph.D. thesis. The data phases obtained from the 1981 experiment have been added to Bonometti's figure.

Whether the compact component is the third image of 0957+561 or the core of the intervening galaxy is still not certain. The change of relative strength of G' with time and wavelength might be the result of "microlensing" action if G' is indeed the third image. At the distance of the galaxy, the Einstein ring radius is 0.03 mas for an object of 100 solar masses. A 1-mJy source of this size corresponds to 10^{11} K at 8.4 GHz, and thus the flux density of a source with this high brightness (which is typical for quasars) could be significantly influenced by individual stars. With a transverse velocity of $0.1c$, the time scale for such variations would be about six months. Microlensing would also produce spectral index differences between G' and A, owing to the spectral index difference between the core and the jet components.

Figure 3 shows the phases on the most sensitive Goldstone-Madrid baseline for the 1981 experiment along with 1983 experiment data and lens model from the work of Bonometti. The 1981 phases are not only consistent with the third image predicted from the lens model but also clearly show the expected trend with an SNR of about 3. If the core of the third image was enhanced in 1983 by microlensing, then the phase variations produced by the image of the jet would be reduced in the 1983 data.

Conclusion. The present maps of the third compact VLBI component in 0957+561 have inadequate SNR and better maps are needed to be completely certain that we have detected the third image along with proof that microlensing is taking place.

Acknowledgment. This paper is an update of the work of Gorenstein *et al.* (1984, 1988). The study of 0957+561 has been vigorously promoted by Professor B. F. Burke and largely funded by the National Science Foundation (NSF). Astronomy at the Haystack Observatory is supported by NSF Grant No. AST-8512598.

References

Bonometti, R.J. 1985, Ph.D. Thesis, Mass. Institute of Technology.

Gorenstein, M. V., *et al.* 1983, *Science*, **19**, 54.

Gorenstein, M. V., *et al.* 1984, *Ap. J.*, **287**, 538.

Gorenstein, M. V., *et al.* 1988, in *IAU Symposium No. 129*, ed. M. J. Reid and J. M. Moran (Dordrecht: Reidel), p. 201.

FIRST VLBI HYBRID MAPS OF 0957+561 A AND B

R. W. Porcas[1], M. Garrett[2], A. Quirrenbach[1], P. N. Wilkinson[2], and D. Walsh[2]

[1] Max-Planck-Institut für Radioastronomie, Bonn
[2] Nuffield Radio Astronomy Laboratories, Jodrell Bank

Very soon after the discovery of the gravitational lens system 0957+561A,B (Walsh, Carswell, and Weymann 1979), VLBI observations using the European VLBI Network (EVN) at 1.6 GHz revealed compact structure in the two image radio components (Porcas *et al.* 1979). More extensive observations in February 1980 enabled the basic core-jet structure of the A and B images to be discovered (Porcas *et al.* 1981). Model-fitting of elliptical Gaussian components to the VLBI visibility functions resulted in parameters for the core-jet separations and position angles, which have been widely used to constrain models of the lensing system. All subsequent published VLBI observations have been analyzed using the model-fitting technique.

Some seven years later (February 1987), we obtained improved 1.6 GHz VLBI data, using the full EVN (Effelsberg, Westerbork, Onsala, Jodrell Bank, Medicina) and Mark III mode B (28 MHz) recording. These data have a factor 4 improvement in sensitivity, better angular resolution and comprise 10 (instead of 3!) baselines. With these data, we have made hybrid maps for both the A and B images of 0957+561 with $\sim$ 18 mas resolution. They are shown in Figure 1.

Although these maps are only preliminary, they give very good confirmation of our previous model-fit results. Measurements from the new data yield very similar values for the core-jet parameters, which are summarized in the table below.

Parameters of 0957+561

	1980 Model	1987 Map
A core-jet separation	46 ± 1 mas	50 mas
A core-jet position angle	$21° \pm 1°$	19.5°
B core-jet separation	56 ± 2 mas	59 mas
B core-jet position angle	$17° \pm 1°$	15.7°

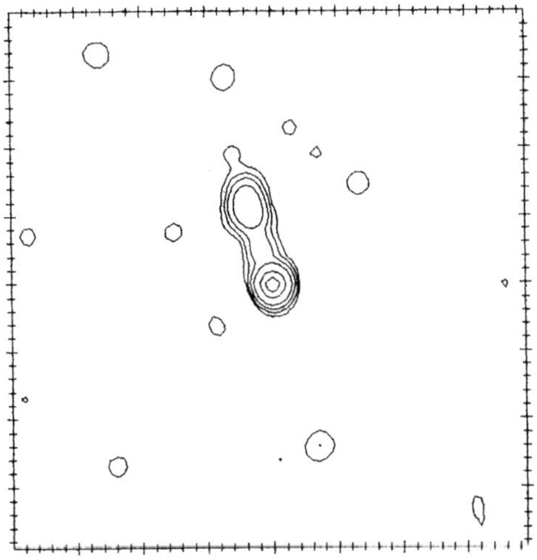

Fig. 1. Hybrid maps of 0957+561 A (left) and B (right) obtained with EVN at 1.6 GHz in 1987. Tick interval = 10 mas; contour levels 2, 5, 10, 20, 50, 80% of the respective peaks.

An easily obtained result from the new data is that both jets point to the northeast (this conclusion was obtained from the older data but involved a tricky argument!).

A new result, which we could not obtain before, is that the core/jet flux density ratios are the same (~ 1.0 at 1.6 GHz) for both A and B. Also, there is little, if any, extended, low brightness emission beyond a radius of about 50 mas. Further analysis of this data set is continuing.

References.

Porcas, R. W., Booth, R. S., Browne, I. W. A., Walsh, D., and Wilkinson, P. N. 1979, *Nature*, **282**, 385.

Porcas, R. W., Booth, R. S., Browne, I. W. A., Walsh, D., and Wilkinson, P. N. 1981, *Nature*, **289**, 758.

Walsh, D., Carswell, R. F., and Weymann, R. J. 1979, *Nature*, **279**, 381.

VLA MEASUREMENT OF THE TIME DELAY IN THE GRAVITATIONALLY LENSED DOUBLE QUASAR 0957+561

J. Lehár,[1] J. N. Hewitt,[2] and D. H. Roberts[3]

[1]Department of Physics 26-335, MIT, Cambridge, MA 02139
[2]Haystack Observatory, Westford, MA 01886
[3]Department of Physics, Brandeis University, Waltham, MA 02254

Abstract. Gravitational lens models for 0957+561 predict that there should be a time delay of roughly a year between the arrival of the two images. We have been monitoring the flux densities of 0957+561 A and B with the VLA for the past eight years. Here we present preliminary light curves of the quasar images constructed from these data. Both images show slow variation, steadily decreasing in brightness for the first few years, then gradually turning upward. We estimate that the time delay is between 400 and 600 days (A leading), and that the magnification ratio (A/B) of the two images is 1.45 ± 0.02.

Introduction. The double quasar 0957+561 is the most securely established and extensively studied candidate for gravitational lensing (Walsh, Carswell, and Weymann 1979; Young *et al.* 1980, 1981; Stockton 1980; Greenfield, Roberts, and Burke 1985; Gorenstein *et al.* 1988). In the lens interpretation, the light that forms the A and B images follows different paths, and we expect there to be a time delay τ between the arrival of the two images (Refsdal 1964); models predict this to be about a year (Young *et al.* 1981). A measurement of this delay would provide further confirmation of the lens hypothesis and allow us to determine the magnification ratio R of the two images. Finally, according to the models of Falco *et al.* (1988), the time delay is related to the Hubble constant H_0 by: $H_0 = (97 \pm 20)(\tau/\mathrm{yr})^{-1}(\sigma_v/390\ \mathrm{km\ s^{-1}})^2 \mathrm{km\ s^{-1}\ Mpc^{-1}}$, where σ_v is the velocity dispersion of the principal lensing galaxy G1.

The Image Light Curves. We have monitored 0957+561 over the past eight years with the Very Large Array (VLA) and have obtained 65 snapshots at 5 GHz (λ 6 cm). Of these, 37 were taken in the more extended configurations of the VLA (A- and B-arrays) and 25 in the more compact C- and D-arrays. Three more were taken in partial configurations, while the VLA was still under construction. The A, B, and partial array data are presented here, and work on the C-array data is still in progress. In labelling the various components of the source, we follow the convention of Roberts *et al.* (1985).

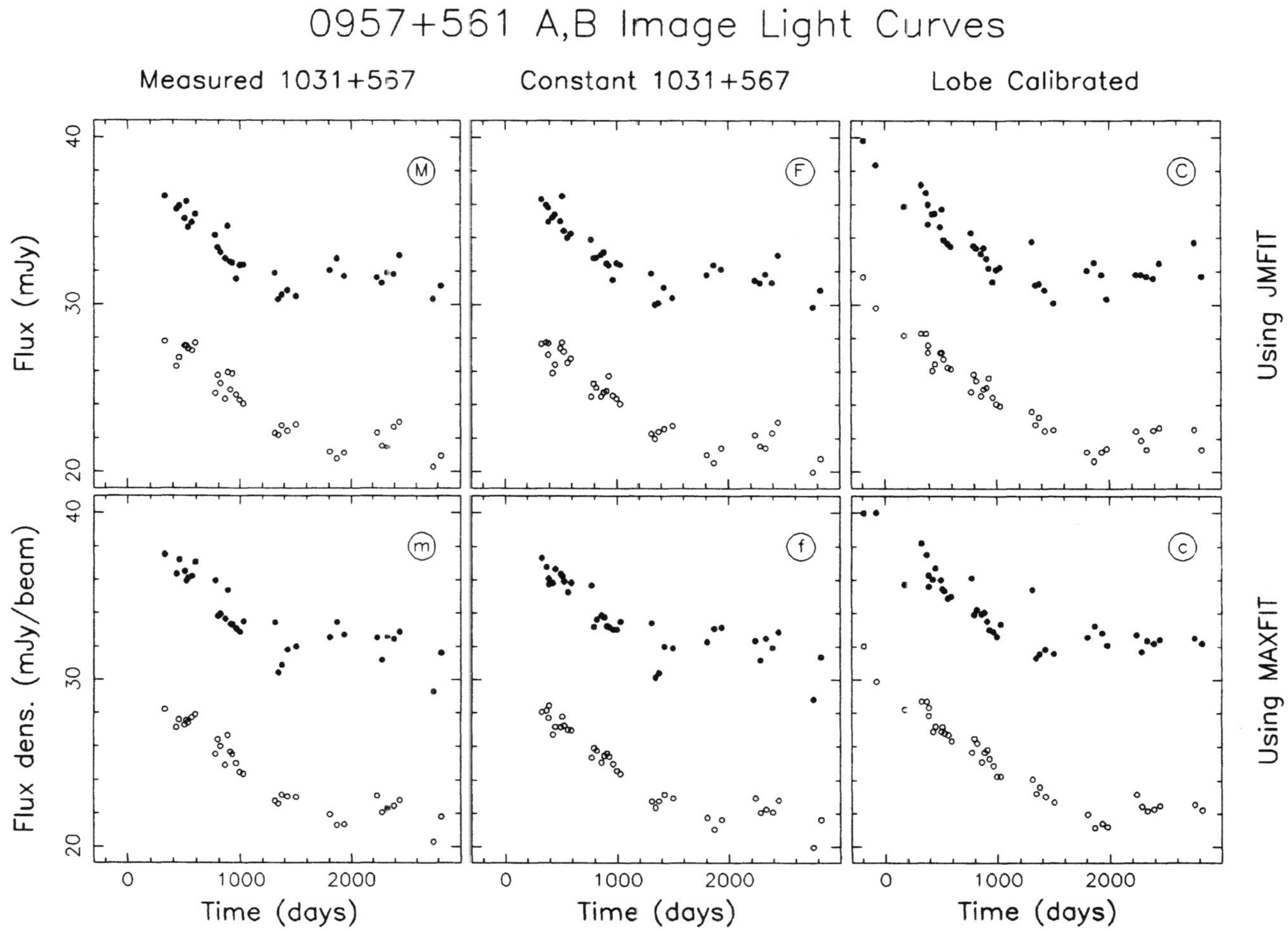

Figure 1: Six light curves for 0957+561 A & B, resulting from 3 calibration and 2 flux density measurement schemes. Filled circles are A image flux densities, empty circles denote B image flux density. Day zero is 1 January 1980. Labels for the curves (used in later figures) are circled.

The individual snapshots were calibrated with respect to 1031+567, and the absolute flux density scale referred to 3C286. Each snapshot was subjected to two iterations of phase self-calibration and mapped in AIPS. The root-mean-square noise in each map is less than 0.2 mJy. Extracting the light curves of the two images proved to be difficult. In the A-array maps, the A and B images are well separated from the surrounding structure; but in the B-array maps, they are contaminated by the jet and G. Two approaches were used to measure the flux density of each quasar image. In the first, we applied a two-component Gaussian fit to a small region around each image, while in the second, the map was convolved to 1$''\!.$5 resolution and the maximum flux density near each image was recorded. We will denote these two approaches by JMFIT and MAXFIT, respectively, referring to the AIPS routines involved. A further complication concerns the flux density scale. The calibrator 1031+567 was found to be constant (within the measurement error) over the entire set of observations, so it may be more appropriate to assume a constant calibrator flux density, rather than use the measured values. Additionally, we produced light curves using the peak flux density of the exterior lobe C, as represented in the 1$''\!.$5 convolved maps, as a reference. The lobe structure is complicated, however, and its peak flux density may be sensitive to baseline coverage. The six sets of light curves (two measurement and three calibration schemes) are displayed in Figure 1. To estimate the uncertainties due to calibration and image flux density extraction, we have subtracted the average of the six light curves from each, and the resulting distributions for the A and B image flux densities are shown in Figure 2. The dispersion about the mean is about 1 mJy for the A image and somewhat less for B.

The Time Delay. In the lens interpretation, the A and B image light curves should differ only by a magnification ratio and a time delay. Their light curves can be qualitatively compared by superimposing a plot of the logarithm of B flux vs. epoch on a similar plot for A. Reasonable correspondence occurs for time delays of about 500 days and magnification ratios of about 1.5. A quantitative determination of τ and R is hindered by the irregular spacing of the snapshots on the time domain. In order to interpolate between measurements, we fit the light curves for both images simultaneously to a single polynomial. For a given τ and R, the B image flux densities were shifted by τ and scaled by R, and combined with the A image data. A polynomial was fitted to the combined light curve, and χ^2 about the polynomial was computed (assuming errors of 1 mJy for each data point). This was done for all six sets of light curves for polynomials of order 2 through 7; the best-fit (τ,R) are shown in Figures 3a and 3b. From their distribution, it seems that the choice of calibration scheme is the largest source of systematic error. An estimate of the time delay and magnification ratio was made by taking the average of the best-fit (τ, R),

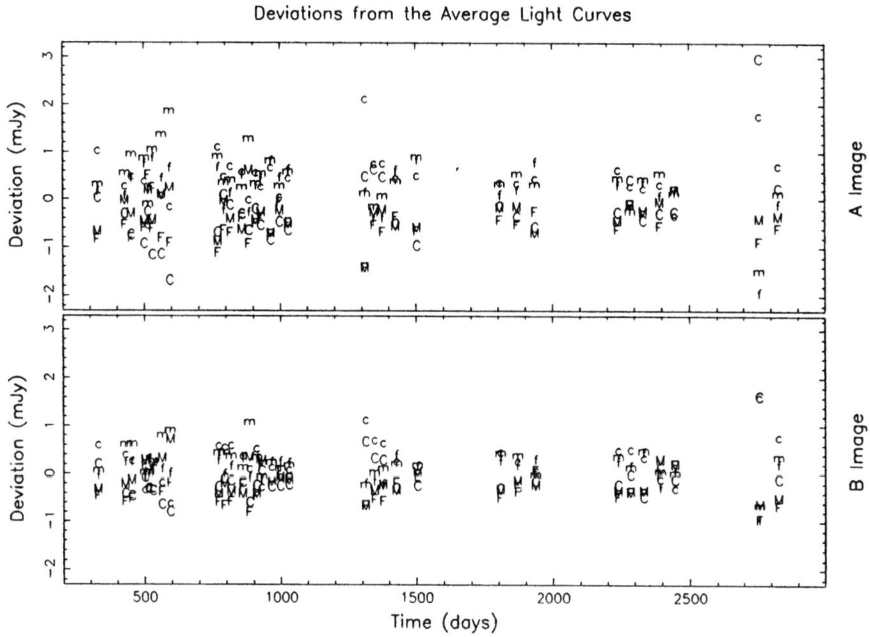

Figure 2: Deviations of the six light curves from their average. See Fig. 1 for labels. Standard deviations: $\sigma_A = 0.672$, $\sigma_B = 0.420$ mJy.

and the tentative uncertainties were taken to be the half-widths of their distribution. This method gives $\tau = 480 \pm 100$ days and $R = 1.45 \pm 0.02$. The fit is illustrated in Figure 4, where a B image light curve was shifted by these amounts and plotted with the A image light curve.

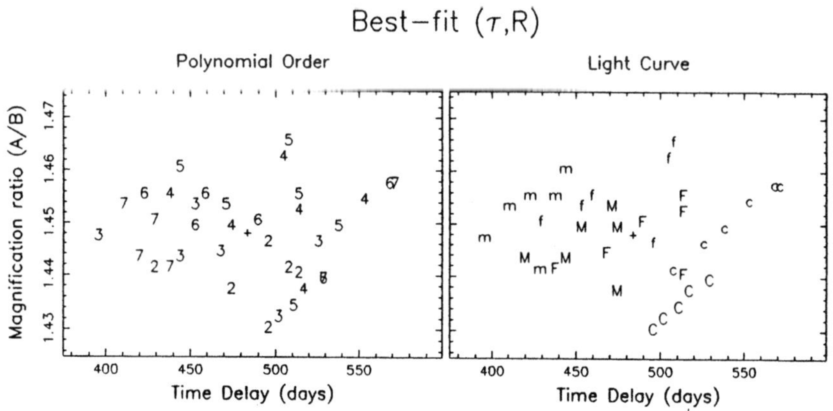

Figure 3: Best-fit (τ, R) from fitting polynomials to the combined A plus scaled & shifted B image curves. The plus sign marks the mean $(\tau, R) = (483^d, 1.448)$. (left) The numbers denote the order of the fitted polynomial. (right) The letters refer to the light curves used (see Fig. 1).

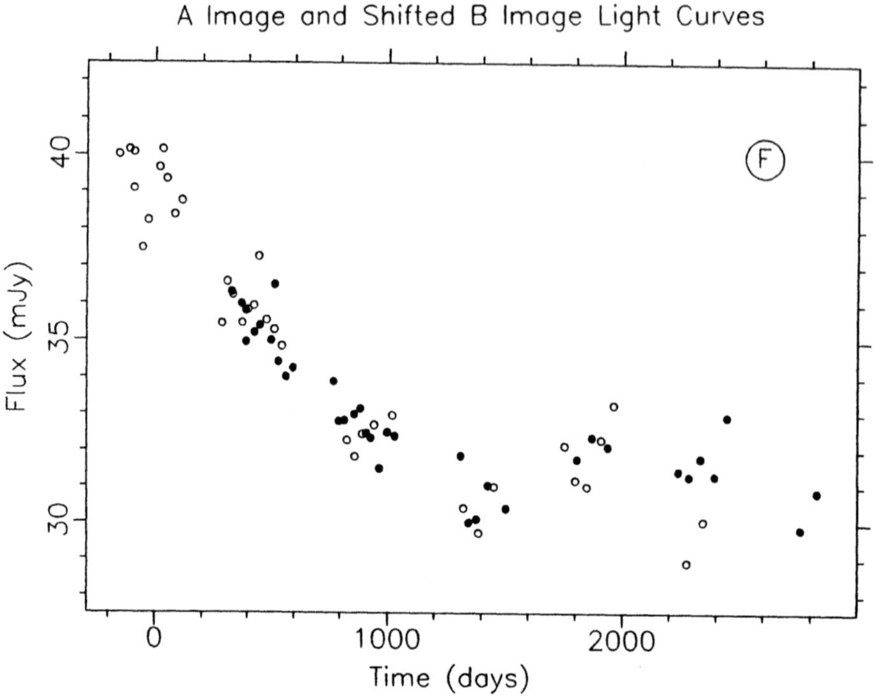

Figure 4: A image plus scaled & shifted B image light curve, using $(\Delta\tau, R) = (483^d, 1.448)$. The fixed 1031+567, JMFIT curves were used. Filled circles are A, empty circles are B.

Discussion. Our time delay of 400 to 600 days includes most of the other estimates (Florentin-Nielsen 1984; Schild and Cholfin 1986; Gondhalekhar *et al.* 1986, Vanderreist *et al.* 1988, 1989). Their values for the magnification ratio, on the other hand, are consistently closer to unity. This could be the result of microlensing, where the B image is amplified by stars as it passes through G1 (Young 1980). As the region of radio emission in the imaged quasar is larger than the optical and ultraviolet emitting regions, we would expect this effect to be weaker in the radio, leading to a larger value of R in the radio. In this case, the time scale for microlensing events must be many years. This could also result from differences between the radio and optically emitting regions. Applying the equation of Falco *et al.* (1988) constrains H_0 to be between 60 and 90 km s^{-1} Mpc^{-1}. While this result is not revolutionary, it is consistent with previously determined values for H_0.

Our estimate for τ would be considerably improved if the large gaps in the time sampling of the light curves were filled. These gaps are due to the fact that for half the time, the VLA is in the C and D configurations. We do have a number of snapshots from the compact arrays, but their low resolution has made image flux density extraction very difficult. We are currently applying an approach similar to that of Masson (1986) to extract the A and B image fluxes from the C-array snapshots. When these and new snapshots are added to revised A-array and B-array flux densities, we expect to arrive at a more precise estimate of the time delay.

We thank B. F. Burke for his many contributions to this work, which has been supported by the NSF under grants at MIT, Brandeis, and Haystack.

References.

Falco, E. E., Gorenstein, M. V., and Shapiro, I. I. 1988, in *The Impact of VLBI on Astrophysics and Geophysics, Proceedings of IAU Symposium No. 129*, ed. M. J. Reid and J. M. Moran (Dordrecht: Reidel), p. 207.

Florentin-Nielsen, R., 1984, *Astr. Ap.*, **138**, L19.

Gondhalekhar, P. M., Wilson, R., Dupree, A. K., and Burke, B. F., 1986, in *New Insights in Astrophysics: Eight Years of Astronomy with IUE, Proceedings of the London Conference*, ESA Publication SP-263.

Gorenstein, M. V., Bonometti, R. J., Cohen, N. L., Falco, E. E., Shapiro, I. I., Bartel, N., Rogers, A. E. E., Marcaide, J. M., Clark, T. A., 1988, in *The Impact of VLBI on Astrophysics and Geophysics, Proceedings of IAU Symposium No. 129*, ed. M. J. Reid and J. M. Moran (Dordrecht: Reidel), p. 201.

Greenfield, P. E., Roberts, D. H., and Burke, B. F. 1985, *Ap. J.*, **138**, 370.

Masson, C. R., 1986, *Ap. J. (Letters)*, **302**, L27.

Refsdal, S., 1964, *M. N. R. A. S.*, **128**, 307.

Roberts, D. H., Greenfield, P. E., Hewitt, J. N., Burke, B. F., and Dupree, A. K., 1985, *Ap. J.*, **293**, 356.

Schild, R. E., and Cholfin, B., 1986, *Ap. J.*, **300**, 209.

Stockton, A., 1980, *Ap. J. (Letters)*, **242**, L141.

Vanderreist, C., Schneider J., Herpe, G., Chevreton, M., Moles, M., and Wlérick, G., 1988, submitted to *Astr. Ap.*

Vanderreist, C., Schneider J., Herpe, G., Chevreton, M., Wlérick, G., and Moles, M. 1989, this volume.

Walsh, D., Carswell, R. F., and Weymann, R. J., 1979, *Nature*, **279**, 381.

Young, P., 1980, *Ap. J.*, **244**, 756.

Young, P., Gunn, J. E., Kristian, J., Oke, J. B., and Westphal, J., 1980, *Ap. J.*, **241**, 507.

Young, P., Gunn, J. E., Kristian, J., Oke, J. B., and Westphal, J., 1981, *Ap. J.*, **244**, 736.

OPTICAL DETERMINATIONS OF THE TIME DELAY IN 0957+561

Christian Vanderriest, Jean Schneider, Georges Herpe, Michel Chevreton, Gérard Wlérick
Observatoire de Paris-Meudon
F-92195 Meudon, Principal Cedex, France
and
Mariano Moles
Instituto de Astrofisica de Andalucia
Apdo. Postal 2144, Granada, 18080, Spain

(presented by G. Soucail)

Summary. A photometric monitoring of the "double" quasar 0957+561 from 1980 to 1987 gives a value for the time delay between the images A and B of $\Delta t_{A,B} \simeq 415 \pm 20$ days with a highly significant correlation peak. A detailed analysis of the data reveals more subtle features:

(1) The true brightness ratio between the two images (measured at the same emission time) is slightly and slowly varying in time.

(2) This brightness ratio $\simeq 1.0$ is different from that obtained for extended components of the quasar at radio wavelengths (VLBI jet) as well as the optical wavelengths (emission lines).

These results suggest the possible presence of a "microlensing" effect on image B.

Introduction. When a massive deflecting body is interposed between a distant source and an observer, the apparent position, brightness, and shape of the image of the source are modified. In our locally inhomogeneous universe, such *gravitational lensing effects* should be common (although not necessarily obvious). Under favorable conditions, the less common but more recognizable phenomenon of *gravitational mirage* can occur, i.e., several distinct images of the source can be seen (for a justification of this terminology, see, e.g., Vanderriest 1985). This extreme case of image distortion has singular properties. Well before the discovery of the first gravitational mirage, it was shown that the complete knowledge of an observed case, including measurement of the travel time delay between the different images, could lead to a direct estimation of H_0 (e.g., Refsdal 1964, 1966, and references therein). However, this determination necessitates a good knowledge of the gravitational potential of the lens.

The "double" quasar 0957+561 (Walsh, Carswell, and Weymann 1979) is almost certainly a gravitational mirage. The main lensing galaxy G1 was detected by Young *et al.* (1980), its structure analyzed by Stockton (1980), and the measurement of its redshift refined more precisely by Young *et al.* (1981). Since then, photometric monitoring has been undertaken by several teams in order to find the time delay $\Delta t_{A,B}$ between the variations of images A and B, but the published results are not yet completely convincing. We present here the yield of our eight-year survey leading to an improved value of $\Delta t_{A,B}$ with a higher degree of confidence. While being the first step on the way to H_0, this measurement of $\Delta t_{A,B}$ is also the definitive proof that A and B are really two images of a single distant quasar.

Sources of Data. Previous observations showed that the variability of 0957+561 at optical wavelengths is small, on the order of 0.1 mag, or less on a time scale of months (Keel 1982; Schild and Weekes 1984), but this variability can also reach several hundredths of a magnitude on a time scale of a few days only (Vanderriest *et al.* 1982; confirmed by Schild and Cholfin 1986). So even with a time sampling as tight as possible, the necessary monitoring could be much longer than $\Delta t_{A,B}$ itself, and a very good photometric precision is required. We thus need a specially dedicated instrument and/or we have to combine different sources of data. Here, our sources were: electronography, photography, photoelectric photometry, and we checked their mutual consistency as well as their agreement with other published data. The observations were made in the standard B band or have been reduced to this color.

a) Electronography. We observed 0957+561 at the 1.93-m telescope of the Observatoire de Haute Provence (O.H.P.) and the 3.6-m Canada-France-Hawaii telescope, with different kinds of cameras (Wlérick 1969, Baudrand *et al.* 1982, Wlérick *et al.* 1983) that have large fields of view (6′ to 10′). This allows us to record several stars from Keel's sequence (1982) whose magnitudes have been improved over the initial IRIS photometer measurements. The data reduction procedure itself is described elsewhere (Vanderriest *et al.* 1982).

On several occasions, observations were made in the U, B, and V Johnson bands during the same night, under good seeing conditions. On these nights, after a careful subtraction of G1, we found that the colors of A and B are essentially the same $(U-B) \simeq -0.82 \pm 0.04$; $(B-V) \simeq 0.13 \pm 0.02$. This result is consistent with the intrinsic colors of the source being constant in time and the reddening from G1 being negligible, as one would expect from an elliptical galaxy. This also ensures that observations in different colors can be safely compared. Thus, although the data obtained by electronography are, by themselves, too sparse for the determination of $\Delta t_{A,B}$, they provide a firm foundation for other observing techniques.

b) Photography. The only bidimensional detector easily available at O.H.P. for routine monitoring was the photographic plate on the 0.6-m aperture Schmidt telescope. Because of the very small scale ($1'' = 10\ \mu$m), we used the combination of hypersensitized IIIaJ emulsion + G.G. 385 and then converted the "J" magnitudes into standard B magnitudes. The data were digitized with a P.D.S. microdensitometer and summed within a 15″ width strip along the A–B axis for the quasar and a dozen calibrating stars. The resulting one-dimensional profiles were fitted by one Gaussian for the stars and by two Gaussians with identical widths for the quasar images. Then, we read directly the magnitudes of A and B by interpolation on the plot (density integral vs. magnitude). This procedure works well because we have several stars with magnitudes close to those of the quasar images. Analyzing the differences between pairs of plates obtained on the same night gives an idea of the mean internal precision; we find $\sigma(A) = 0.03$ and $\sigma(B) = 0.05$ mag. A few plates were also obtained at the 1.2-m telescope (F/6; $1'' = 35\ \mu$m) with baked IIaO emulsion + G.G. 385 filter. They were processed in the same way as the Schmidt plates, except that no color correction is needed. The precision is about 0.04 mag.

c) Photoelectric photometry. In order to obtain frequent data with very good precision, we used scanning photometers (for a description, see Vanderriest *et al.* 1988 and also Chevreton *et al.* 1983) permanently attached to easily accessible telescopes. The first photometer (PAB1) was mounted at the 1.93-m telescope of the O.H.P. and a second one (PAB2) was mounted later at the 1.5-m Spanish telescope of the Calar Alto Observatory.

The observations had to be made on a "time sharing" basis previously unknown to program committees. Ideally, the telescope should be available during two hours for each measurement, three times a month, eight months per year, . . ., for an undetermined number of years!

The output data are directly in the form of one-dimensional profiles of the quasar images and of a single reference star: S5 from Keel's sequence, 63″ away from the quasar. The data processing again consists of Gaussian fitting and subtraction of G1. When the seeing was better than 2.″5, the expected precision of 0.02 mag was effectively reached.

d) Published C.C.D. data from Schild and collaborators. Because they seemed of very good quality and would help to extend the time coverage, we tried to add the data published by Schild *et al.* (Schild and Weekes 1984; Schild and Cholfin 1986) to our own data set. For that, we had to transform their R magnitudes into B magnitudes, i.e., we needed to know precisely the $(B - R)$ index of the quasar. By chance, some of our B observations were almost simultaneous with Schild's observations on several occasions. We found that $(B - R)$ was remarkably constant for each image, as with the $(U - B)$ and $(B - V)$ indices obtained by electronography. But, this time, there was a significant difference between the two images: $\langle \text{B-R} \rangle = 0.90 \pm 0.02$ for A and $\langle \text{B-R} \rangle = 0.82 \pm 0.02$ for B. A detailed analysis (Vanderriest *et al.* 1988) showed that the most likely cause of this discrepancy was an overestimation by Schild and collaborators of the flux of G1. Moreover, they noted a probably zero error of 0.12 mag in their R magnitude scale. The true $(B - R)$ index of the quasar thus becomes 0.78, in fair agreement with the direct measurements by Beskin, Neizvestnyi, and Shvartsman (1980) for A, the less contaminated image. In any case, the corrected data are fully usable and have been associated with ours.

e) Other data. The scatter remains high between our data and those of Keel (1982). This scatter probably results from the limited intrinsic precision of the IRIS photometer method and also from the rather large errors on his photometric sequence (about 0.15 mag at magnitudes comparable to those of the quasar images).

The data of Florentin-Nielsen (1984) were also too noisy and were not used.

Thus, our starting data set consists of 131 good measurements, carefully reduced to the same photometric system: 12 data points were obtained by electronography; 40 by photography (28 at the Schmidt telescope and 12 at the 1.2-m telescope); 50 by scanning photometers (45 for PAB1 and 5 for PAB2); and 29 from the C.C.D. data published by Schild and collaborators.

Data Analysis; The Value of $\Delta t_{A,B}$. The entire data set is plotted in Figure 1. The long-term tendancy is towards a slight brightening of both A and B, but a sharp 0.25 mag decline is also visible between February and May 1984 for A and between April and July 1985 for B. This feature, already suspected by Schild (1986), suggests a possible $\Delta t_{A,B}$ around 400 days.

a) Theoretical and practical problems. The observed fluxes of images A and B should be related by:

$$B(t + \Delta t_{A,B}) = R \cdot A(t) \quad ,$$

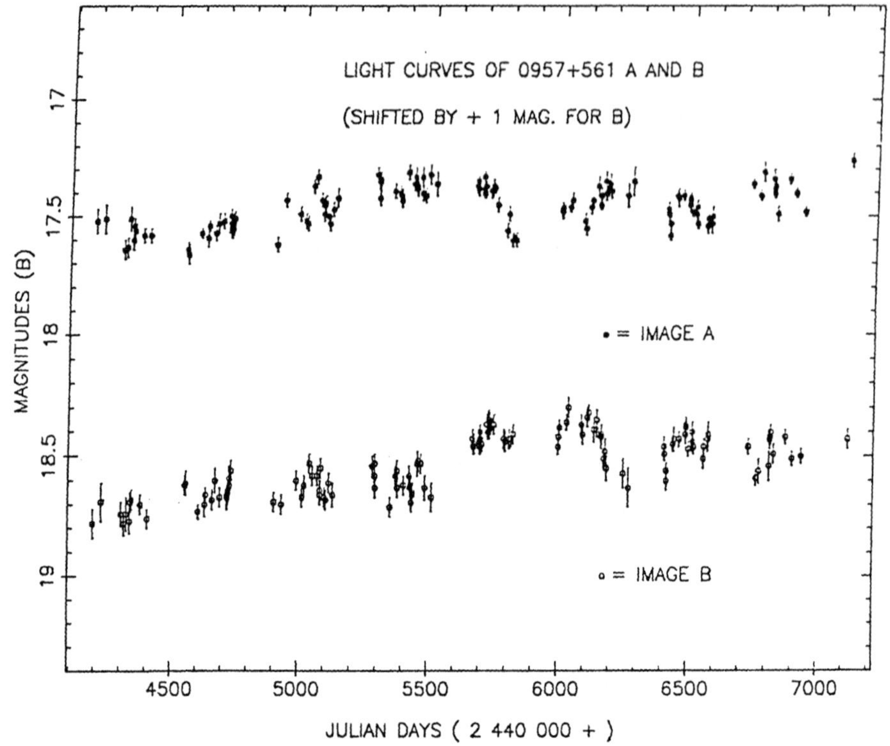

Figure 1. Observed light curves.

with $R = \mathcal{M}_B/\mathcal{M}_A$, the ratio of the magnification factors for images A and B. The knowledge of R is necessary for the model, but, because of the variations of the relative lens-source position, R is also a time-dependent parameter. The variation of the smoothed out potential of the lens galaxy is very slow and practically negligible. On the other hand, we should subtract the microlensing effects due to individual stars of this galaxy crossing the light beam.

A possibility could be to determine $\Delta t_{A,B}$ from the variations of some compact part of the source (for the present work: optical continuum) and R from the flux ratio of a sufficiently extended component. By "sufficiently extended," we mean a component that is large enough to be insensitive to the microlensing effect of individual stars but small enough with respect to the scale of variation of the galaxy potential. Such a component of the source could also be considered over a time of the order of $\Delta t_{A,B}$ and R would be simply $B(t)/A(t)$ instead of $B(t + \Delta t_{A,B})/A(t)$. We will discuss later two possible candidates: 1) the VLBI jet, and 2) the emission line region (cf. Rees 1981).

b) Determination of $\Delta t_{A,B}$. Another difficulty comes from the sampling of the light curves. Starting from the data point $A(t_i)$ and $B(t_i)$ at observing times t_i, we create the continuous light curves $A(t)$ and $B(t)$ by interpolation and calculate the correlation coeffient $C(\Delta t)$ by a standard cross-correlation procedure.

Figure 2 shows, for the whole data set, a maximum at $\Delta t_{A,B} = 419 \pm 30$ days with $C(419) = 0.62$. If we restrict the correlation to the period 1983–1986, which shows the most prominent feature, the correlation coefficient reaches almost 0.8. On the other hand, the procedure applied to subsamples (photographic data only, or photoelectric, or Schild's) gives values of $\Delta t_{A,B}$ in the range 390 to 440 days with less significant correlation coefficients (0.2 to 0.5).

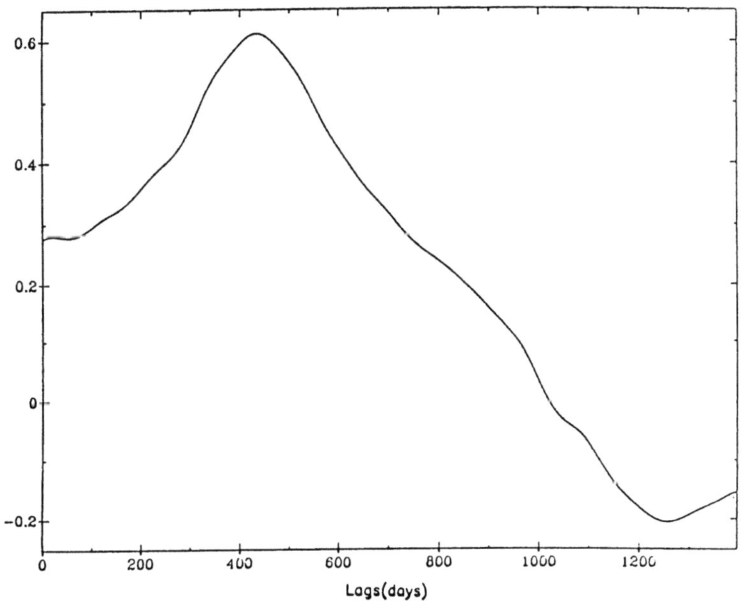

Figure 2. Cross-correlation function.

The method of discrete Fourier transform is, theoretically, more suited to irregularly sampled data. A classical algorithm (Moles *et al.* 1986) gives $\Delta t_{A,B} = 412 \pm 30$ days, in very good agreement with the result from interpolated data. We adopt the mean value $\Delta t_{A,B} = 415 \pm 20$ days.

c) Possible microlensing effect. Figure 3 shows the superposition of the data for the optimum values $\Delta t_{A,B} = 415$ days and $\langle B\rangle/\langle A\rangle = 0.97$. Although the overall agreement is good, a slightly different value of B/A could have been preferred in some parts of the plot. A study of the variation of $R(t) = B(t + \Delta t_{A,B})/A(t)$ would help to quantify this impression, but, again, we are facing a problem of time sampling. As the effect is small, a correct analsis cannot be done on the *interpolated* data. On the other hand, there are very few observation points $A(t_i)$ and $B(t_j)$ that satisfy exactly the condition $(t_j - t_i) = \Delta t_{A,B}$. So, we can relax the condition and consider all pairs of points satisfying the condition $(t_j - t_i) = \Delta t + dt$, with dt being in the "window" $[-\epsilon, +\epsilon]$. Because of the rapid intrinsic variations of the quasar already mentioned, ϵ should not be too large, but it should not be too small to have an insufficient number of pairs. In Figure 4, we took $\epsilon = 5$ days. For the polynomial fitting, each point has been weighted according, first, to the measurement errors on A and B and, second, to the more or less close match to $\Delta t_{A,B}$ (the weighting function was taken, quite arbitrarily, as $e^{-dt/\tau}$, with $\tau = 10$ days). The residue to a polynomial fit (degree 3) is $\sigma_3 = 0.054$, close to the precision of the starting data, while it is $\sigma_0 = 0.067$ for the fit by a constant. A statistical analysis shows that this improvement is significant at the 3σ level. So we conclude cautiously that variations of the amplification ratio have been marginally observed from direct inspection of $R(t)$.

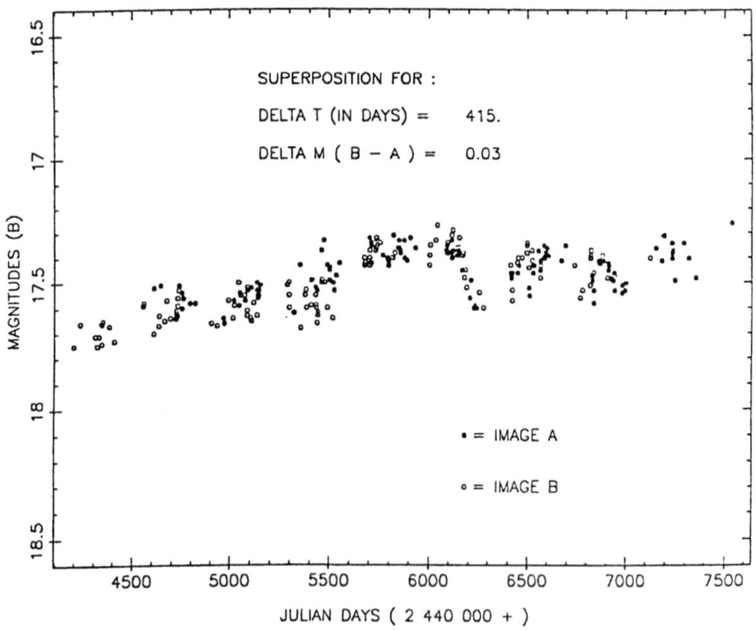

Figure 3. Superposition of the two light curves with a time delay of 415 days.

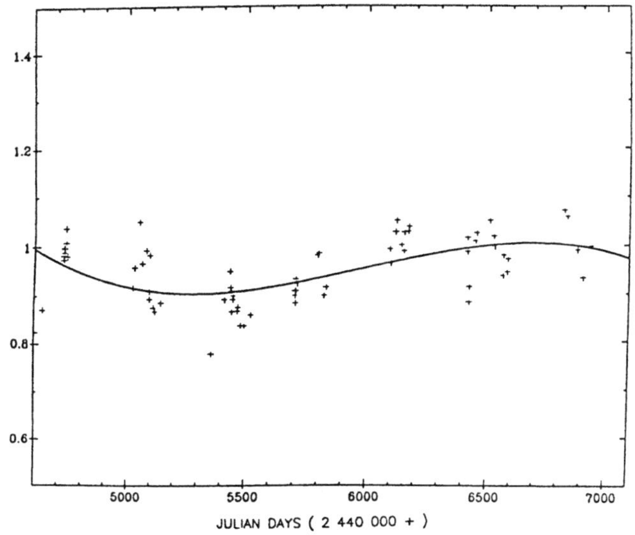

Figure 4. A/B ratio reduced at the same emission time (delay 415 days) and polynomial fit.

A firmer clue, perhaps in favor of the presence of microlensing is the value of R itself, found to be about 1. Compared to the values for extended components, i.e., $\simeq 0.73$ for the optical emission lines (Wills and Wills 1980; Young *et al.* 1981) or $\simeq 0.64$ from the VLBI jet (components 2 and 3; Gorenstein *et al.* 1988a), this indicates a possible microlensing enhancement of ≈ 0.3 mag amplitude for image B. Such small amplitude and long duration events appear when a source is only approaching the caustic of a star and not actually crossing it (Young 1981; Kayser, Refsdal, and Stabell 1986). They have a good probability of being observed for image B, which is seen through rather dense parts of G1.

Conclusion.

1) The unambiguous determination of $\Delta t_{A,B}$ confirms the validity of the gravitational mirage theory and could be a starting point for further developments. The precision on its value could still be improved (to, perhaps, ± 1 day) by intensive monitoring during two short periods of time separated by 415 days.

2) Borgeest (1986) has shown that, in first approximation, the mass of G1 can be deduced from $\Delta t_{A,B}$ independently of the surrounding cluster. With our value of $\Delta t_{A,B}$, we obtain $M(G1) = 8.7 \times 10^{11}\ M_{\odot}$ within a $3\farcs1$ radius, and, with our photometry of G1, $(M/L)_V \simeq 25h^2$ ($h = \mathrm{H_0}/100$ km s^{-1} Mpc^{-1}), a reasonable value for an elliptical galaxy.

3) On the other hand, the knowledge of $\Delta t_{A,B}$ and of the characteristics of two images is not sufficient for finding a secure value of H_0 (Gorenstein *et al.* 1988b), but the degeneracy could be broken by observation of the third image and of the structure of the deflector.

Current models do not lead to absurd results but also do not put stringent constraints on H_0. With our value of $\Delta t_{A,B}$, the model of Borgeest and Refsdal (1984) gives a firm upper limit $H_0 < 175$ and a "best" estimate $H_0 \simeq 105$. This can be compared to the limit obtained by Falco (1987) from VLBI data only: $H_0 < 140$. Another model of the galaxy-cluster deflector (Falco, Gorenstein, and Shapiro 1985) would give $H_0 \leq 86$. There is a clear need for additional constraints from the observations, first of all a better modeling of the associated cluster, by systematic redshift measurements in the field of 0957+561. If several gravitational mirages could be analyzed and could give independent estimations of H_0, the influence of background inhomogeneities (Alcock and Anderson 1985) could be tested.

4) The microlensing issue deserves further study. The amplification ratio $R = \langle B \rangle / \langle A \rangle$ due to the "smooth" potential of the lens is an important parameter for the model. A value, free of microlensing effects, about $R = 0.7$, is suggested by the observation of extended parts (optical emission lines, VLBI jets), but this should be confirmed by better VLBI and spectrophotometric observations. The compact parts (optical continuum, compact radio core) are probably affected by microlensing. If this is true and if their sizes are comparable, the delayed ratio $B(t + \Delta t_{A,B})/A(t)$ should be the same for these radio and optical components. This could be tested by two-epoch observations.

References.

Alcock, C., and Anderson, N. 1985, *Ap. J. (Letters)*, **291**, L29.

Baudrand, J., Chevillot, A., Dupin, J. P., Bellenger, R., Félenbok, P., Guérin, J., Picat, J. P., and Vanderriest, C. 1982, *J. Optics Paris*, **13**, 295.

Beskin, G., Neizvestnyi, S., and Shvartsman, V. 1980, IAU Circular 3533.

Borgeest, U. 1986, *Ap. J.*, **309**, 467.

Borgeest, U., and Refsdal, S. 1984, *Astr. Ap.*, **141**, 318.

Chevreton, M, *et al.* 1983, 24th Liège Astrophysical Colloquium, p. 161.

Falco, E. 1987, *Rev. Mexic. Astr. Ap.*, **14**, 127.

Falco, E., Gorenstein, M., and Shapiro, I. 1985, *Ap. J. (Letters)*, **289**, L1.

Florentin-Nielsen, R. 1984, *Astr. Ap.*, **138**, L19.

Gorenstein, M., Bonometti, R., Cohen, N., Falco, E., Shapiro, I., Bartel, N., Rogers, A., Marcaide, J., and Clark, T. 1988a, in *The Impact of VLBI on Astrophysics and Geophysics, Proceedings of IAU Symposium No. 129*, ed. M. J. Reid and J. M. Moran (Dordrecht: Reidel), p. 201.

Gorenstein, M., Falco, E., and Shapiro, I. 1988b, *Ap. J.*, in press.

Kayser, R., Refsdal, S., and Stabel, R. 1986, *Astr. Ap.*, **166**, 36.

Keel, W. 1982, *Ap. J.*, **255**, 20.

Moles, M., Garcia-Pelayo, J. M., Masegosa, J., and Garrido, R. 1986, *A. J.*, **92**, 1030.

Rees, M. 1981, in *ESO Conference on "Scientific Importance of High Angular Resolution at Infrared and Optical Wavelengths*, ed. M. H. Ulrich and K. Kjär, p. 423.

Refsdal, S. 1964, *M. N. R. A. S.*, **128**, 295 and 307.

Refsdal, S. 1966, *M. N. R. A. S.*, **132**, 101.

Schild, R. 1986, in *Quasars, Proceedings of IAU Symposium No. 119*, ed. G. Swarup and V. K. Kapahi (Dordrecht: Reidel), p. 549.

Schild, R., and Cholfin, B. 1986, *Ap. J.*, **300**, 209.

Schild, R., and Weekes, T. 1984, *Ap. J.*, **277**, 481.

Stockton, A. 1980, *Ap. J. (Letters)*, **242**, L141.

Vanderriest, C. 1985, in *Lecture Notes in Physics*, **212**, 265.

Vanderriest, C., Bijaoui, A., Félenbok, P., Lelièvre, G., Schneider, J., and Wlérick, G. 1982, *Astr. Ap.*, **110**, L11.

Vanderriest, C., Schneider, J., Herpe, G., Chevreton, M., Wlérick, G., and Moles, M. 1988, submitted to *Astr. Ap.*

Walsh, D., Carswell, R., and Weymann, R. 1979, *Nature*, **279**, 381.

Wills, B., and Wills, D. 1980, *Ap. J.*, **238**, 1.

Wlérick, G. 1969, *Adv. E. E. P.*, **28B**, 787.

Wlérick, G., Lelièvre, G., Servan, B., Cayatte, V., Michet, D., Renard, L., and Horville, D. 1983, in *Instrumentation in Astronomy V*, ed. A. Bocksenberg and D. Crawford, p. 143.

Young, P. 1981, *Ap. J.*, **244**, 756.

Young, P., Gunn, J., Kristian, J., Oke, J., and Westphal, J. 1980, *Ap. J.*, **241**, 507.

Young, P., Gunn, J., Kristian, J., Oke, J., and Westphal, J. 1981, *Ap. J.*, **244**, 736.

RESOLUTION OF GALAXY AND THIRD IMAGE OF GRAVITATIONAL LENS 2016+112

G. Langston[1], C. Carilli, S. Conner, M. Heflin, J. Lehár,
C. Lawrence[2], V. Dhawan[3], and B. Burke
Massachusetts Institute of Technology
Cambridge, MA 02139 USA

Introduction. We present $\lambda\lambda$ 2, 6 and 90 cm VLA observations of the gravitationally lensed quasar 2016+112. The radio image of lens system 2016+112 is an unusual collection of three objects, A, B and C, arranged in a right triangle. Objects A and B are two images of a single high redshift quasar. Component C is coincident with a lower redshift galaxy. Narasimha, Subramanian, and Chitre (1987) have produced models estimating the pathlength difference between the two images A and B as well as predicting a brightness of a third image of the quasar near C. These observations were made to determine structure of of radio component C and to begin monitoring the flux density variations of images A and B. Flux density monitoring is the first step in measuring the difference in the two image path lengths.

Figure 1 shows the location of the images and lenses in the λ 2 cm map produced from these new observations. Lawrence *et al.* (1984) showed that components A and B have nearly identical optical spectra with narrow, high redshift emission lines. The spectral index, α, between $\lambda\lambda$ 6 and 20 cm is similar for A and B, $\alpha = 0.8 \pm 0.06$ ($S \propto \nu^{-\alpha}$). The spectral index of C is flatter, $\alpha = 0.24 \pm 0.06$. The lens model predicts that A and B will have similar spectral indices at all frequencies and these observations check the spectral index between $\lambda\lambda$ 2 and 6 cm. Component C is three times brighter than both A and B at 6 cm but fainter at optical wavelengths. Schneider *et al.* (1986) found high redshift Lyman α emission near C that is five times fainter than component A. Component C is apparently composite, but the relative contribution of the quasar image and the foreground galaxy was not determined. Following Schneider *et al.* (1986), the image of the quasar near the galaxy will be called C'. The flux from the galaxy will be called C. Optical observations also show one large galaxy, D, south of a line connecting images A and B. Schneider *et al.* (1986) found the galaxy D had a redshift of 1.01 ± 0.005.

---

[1] Current Address: Max Planck Institut, Auf dem Hügel 69, D-5300 Bonn-1, FRG
[2] Current Address: California Institute of Technology, Pasadena, California USA
[3] Current Address: Harvard Smithsonian Obs., 60 Garden St, Cambridge MA 02139 USA

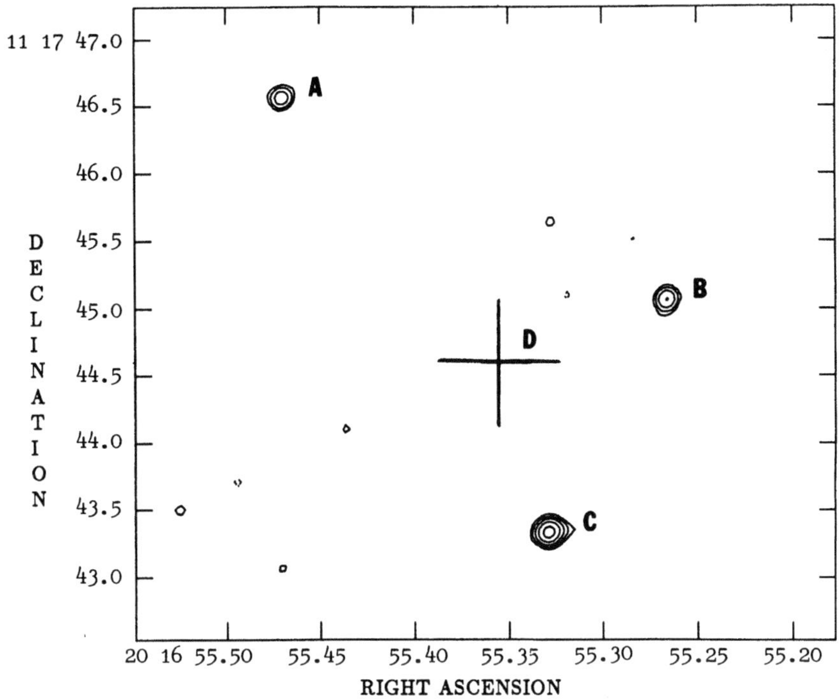

Fig. 1. Radio map of 2016+112 at 2 cm with 0.12 arcsecond resolution. Galaxy D is marked with 0.5 arcsecond position error bars.

Observations. The new observations presented here were made in July 1987 with the VLA in the A configuration. Images A and B are not resolved in the λ 2 cm VLA map with Gaussian beam size of 0.1216×0.1173 arcseconds and position angle of -15 degrees East from North. Component C is resolved. A single Gaussian fit has major and minor axes of 0.1398 ± 0.0019 and 0.1202 ± 0.0019 arcseconds, respectively, with position angle of 101 degrees. The 1.16 major to minor axis ratio of the fit is significantly greater than the 1.04 axis ratio of the beam.

The total 6 cm flux density of the three components was observed to decrease relative to flux density observed in February 1982 by Lawrence *et al.* (1984). The flux density observed in May 1984 by Schneider *et al.* (1986) is consistent with a continuous decrease in flux density between 1982 and 1987. The May 1984 observations of individual component flux densities have larger errors because the components are not resolved. In February 1982 the flux was 113 mJy, in May 1984, 108 mJy, and in July 1987 the flux density was 101 mJy. This is an overall decrease of 12% in 5.3 years. The flux densities of the two images, A and B, are best compared relative to the flux density of C at each epoch. This method avoids the problem of determining the absolute flux density scale over an extended period. Table 1 shows the variation in the flux densities of A and B relative to C. Figure 2 presents the 6 cm flux ratios as a function of time.

Table 1
Flux Density of Components of 2016+112 and Ratios of Images to Component C

Comp.	1982 4.86 GHz (mJy)	1984 4.86 GHz (mJy)	1987 4.84 GHz (mJy)	1987 4.88 GHz (mJy)
A	21.9 ± 0.5	20.5 ± 0.5	19.22 ± 0.15	18.93 ± 0.15
B	23.2 ± 0.5	21.9 ± 0.5	19.82 ± 0.15	20.05 ± 0.15
C	67.9 ± 1.0	65.8 ± 1.0	61.93 ± 0.15	61.72 ± 0.15
		Ratios		
	(%)	(%)	(%)	(%)
A/C	32.3 ± 0.75	31.2 ± 0.76		30.9 ± 0.24
B/C	34.2 ± 0.75	33.3 ± 0.76		32.2 ± 0.24

Note: The flux density measurements for 1987 were from two 50 MHz bands, and the ratios are of the average flux densities for 1987. Flux density errors are RMS estimates of error; the overall flux density scale is accurate to roughly 5 percent. The comparison of the ratios is independent of the overall flux density scale for each epoch.

Note that A and B are decreasing in brightness relative to C, but the total flux density decrease shows that C is also becoming fainter at 6 cm. The limited data do not show significant features that could be used to determine the time delay between the two image paths.

The new observational results are summarized below:

1. Component C is resolved at λ 2 cm. The subtraction of a point source from the peak location of C yields a model dependent flux density for C'. The ratio of A/C' flux density is consistent with the Lyman α flux ratio presented by Schneider *et al.* (1986). Either this is an image or an image upper limit. None can be further than 0.19 arcseconds and brighter than 0.5 mJy at 2 cm. (The image ratio is found using the component A flux density of 3 mJy: $3/0.5 = 6 \approx 2$ magnitudes).
2. The spectral index between $\lambda\lambda$ 2–6 cm for A, B and C are all similar, 1.4 ± 0.1.
3. The flux densities of A, B *and* C are decreasing at 6 cm. The flux densities of A and B are decreasing more rapidly than the flux density of C.
4. There is a peak in the radio spectra between 90 and 20 cm. The radio spectra are shown in Figure 3.

Images and Lenses. The simplest model for a gravitational lens, a point mass, predicts that images will be found on opposite sides of the lens. Optical observations of 2016+112 show there

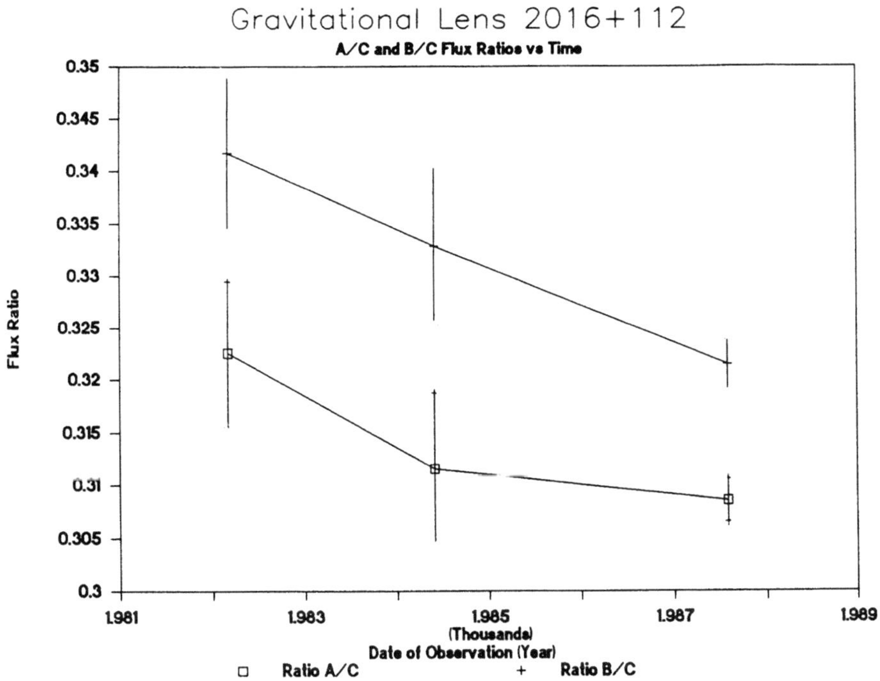

Fig 2. Flux density ratios of A/C and B/C at 6 cm vs. data of observation. The first two observations assume ±0.5 mJy flux density errors. Errors in the absolute flux density scale are not important in this plot as these errors are also reflected in C. The July 1987 errors are measured as 0.15 mJy from separate calibration and processing of two 50 MHz bandpasses.

are two galaxies south of the line between A and B and additional unseen mass is required to explain this offset. Narasimha, Subramanian, and Chitre (1987) have modeled the lens as the two galaxies C and D, plus a massive cluster. Their models predict the radio flux density due to C' and assume the remainder of the radio flux is due to the galaxy C. These observations constrain their models, by placing limits on the separation between C and C'.

Narasimha, Subramanian, and Chitre (1987) produce three models representing the relative importance of the different lensing components. The first of their models is ruled out by these observations because bright quasar images must lie within 0.19 arcseconds of component C. This model represents a lensing scenario with galaxy D dominating. The second of their models is not excluded by these observations, because the A/C' magnification ratio lies below the 2 cm flux density limits. This model places nearly equal masses at C and D. In this case, the mass of C is not represented by the optical light observed. The second model predicts that a quasar image will be seen near C in deeper 2 cm observations. Their third model has a mass distribution dominated by the nucleus of galaxy C. This model has additional images near the galaxy C as

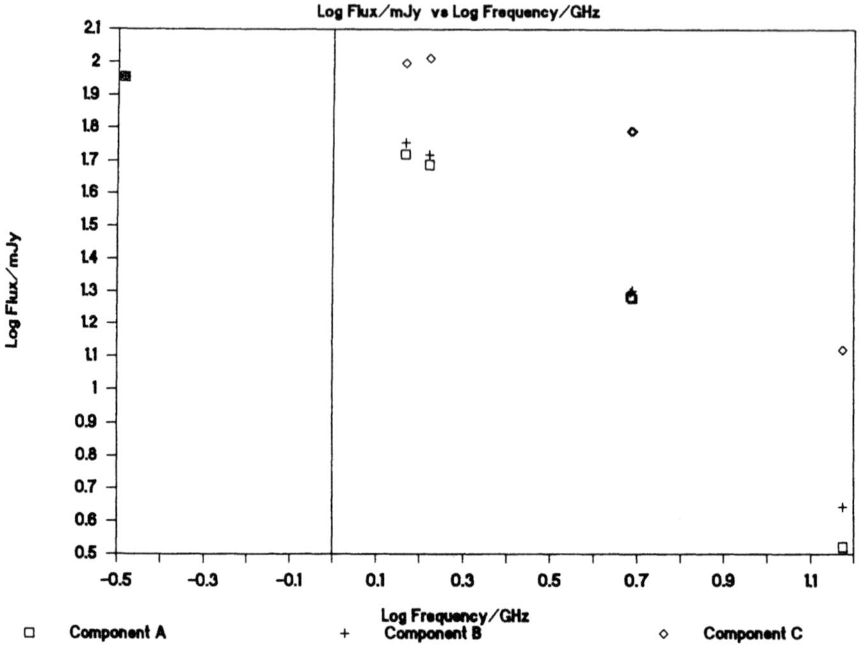

Fig 3. Radio spectrum of 2016+112 A, B, and C. Note that A, B, and C all have a steep spectral index of 1.4 ± 0.1 between 2 and 6 cm. The point at 327 MHz is the integrated flux of A, B, and C combined and is accurate to 20 percent. The separation of the A and B flux density at 14.4 GHz is due to the uncertainty of the measurement.

required by these observations. However, the third model is excluded by the Lyman α flux ratio of A/C' $\approx 1/6$. Their predicted Lyman α flux at C is smaller than that observed. The Lyman α ratio is a lower limit to the ratio of A/C' images, because some obscuration of C' is expected through a galaxy at C. This third type of mass distribution could produce a successful model if a geometry is found that produces a smaller A/C' ratio.

The path length difference is predicted by these models, and further monitoring source variation can be used to measure this path length. Narasimha *et al.*'s models predict the flux density variations at A will lead B by roughly one year. These observations suggest a slow decrease in the flux density of both A and B. This predicts the next observation of B will show a continued decrease.

There must be an image of the quasar, C', very near the galaxy C, to explain the Lyman α emission. Because C and C' have such a small separation, it suggests that all radio flux density could be due entirely to a third quasar image. In this scenario, the galaxy at C obscures the optical Lyman α in order for the radio flux density ratio to be consistent with the Lyman α ratio. This model is contradicted by VLBI data, which indicates that a single image can not contain all the C flux density. Both images A and B are seen in VLBI observations at 18 cm (Heflin *et al.* 1988). A limit is placed on the component C flux density, the compact VLBI image at C has a fainter peak flux density than one quarter the B image peak flux density. The models of Narasimha *et al.* predict several images at C, but if all flux density at C is due to quasar images, there must be many faint images. The difference in optical and radio flux density ratios of C' must be explained by roughly 3 magnitudes of obscuration or variability.

Conclusions. The components of 2016+112 are variable. It is probable that *all* three compact objects are variable, but the absolute flux density scale is difficult to determine. The strong variation of C is contrary to expectations, because VLBI observations show C is extended and must vary more slowly than a compact object. If the C flux density decrease is due to variation of a third image, this image must have very large percentage flux density changes. Any third image, C', must be very near C, < 0.19 arcseconds away for images with A/C' flux density ratios < 2 magnitudes. Component C has been resolved, but measuring the contribution of the quasar image to the radio flux density requires further observations. The Narasimha *et al.* model that is not excluded by these observations suggests that a massive galaxy is located at the radio component C.

References.

Heflin, M. B., Gorenstein, M. V., Falco, E. E., Shapiro, I. I., Burke, B. F., Hewitt, J. N., Rogers, A. E. E., and Lawrence, C. R. 1988, in *The Impact of VLBI on Astrophysics and Geophysics, Proceedings of IAU Symposium No. 129*, ed. M. J. Reid and J. M. Moran (Dordrecht: Reidel), p. 209.

Lawrence, C. R., Schneider, D. P., Schmidt, M., Bennett, C. L., Hewitt, J. N., Burke, B. F., Turner, E. L., and Gunn, J. E. 1984, *Science* **223**, 46.

Narasimha, D., Subramanian, R. and Chitre, S. M. 1987, *Ap. J.*, **315**, 434.

Schneider, D. P., Gunn, J. E., Turner, E. L., Lawrence, C. R., Hewitt, J. N., Schmidt, M., and Burke, B. F. 1986, *Astr. J.*, **91**, 991.

V. Luminous Arcs

ARCS IN CLUSTERS OF GALAXIES AS GRAVITATIONAL LENS IMAGES

Vahé Petrosian
Center for Space Science and Astrophysics
Stanford University, Stanford, CA 94305, USA

Abstract. The observed properties of the arclike images discovered by Lynds and Petrosian in clusters of galaxies are summarized. It is shown that these observations can best be described by gravitational lensing of a distant and somewhat unusual galaxy, by the mass distribution in the clusters. It is further shown how the modeling of the phenomenon can be used to determine the distribution of visible and dark matter in the clusters and to obtain the characteristics of the distant sources.

Introduction. During the spring of 1976 and 1977, Roger Lynds, Allan Sandage, and I obtained multicolor video camera pictures of about 50 large redshift clusters of galaxies with the KPNO 4-meter telescope. The purpose of this program was to obtain the surface brightness distribution of first rank giant elliptical galaxies, which for many years had been considered as suitable candidates for determination of the cosmological parameters via their redshift-magnitude relation. This relation depends not only on the cosmological parameters but also on the evolution of the galaxies, so that the progress in this area has been hampered by the uncertainties about the evolution of these galaxies. The observed surface brightness, b, of a source is independent of the cosmological parameters or distance (in fact, it is independent of the gravitational field or mass distribution along the line of sight) and is related to the intrinsic surface brightness, B, and redshift, z, of the source. The relation for the bolometric brightnesses is $B = 4\pi b(1+z)^4$. Our goal, therefore, was to determine the evolution of galaxies by this procedure (Petrosian 1976) and use this knowledge in cosmological studies.

All clusters with $z \geq 0.2$ known to us at that time were included in this study. We also observed about 20 clusters with redshifts less than 0.2, and for some of them we obtained video camera and prime focus pictures. The redshift distribution of the clusters with $z > 0.1$ is shown in Figure 1. The position of the three clusters, Abell 2218, Abell 370, and a cluster at 2244-02 (Cl 2244, hereafter), with respective redshifts of 0.171, 0.373 and 0.328, which showed unusual filamentary features, are marked on this figure. These data, even though a result of our best effort at that time, could merely indicate to us that these features were extraordinary and

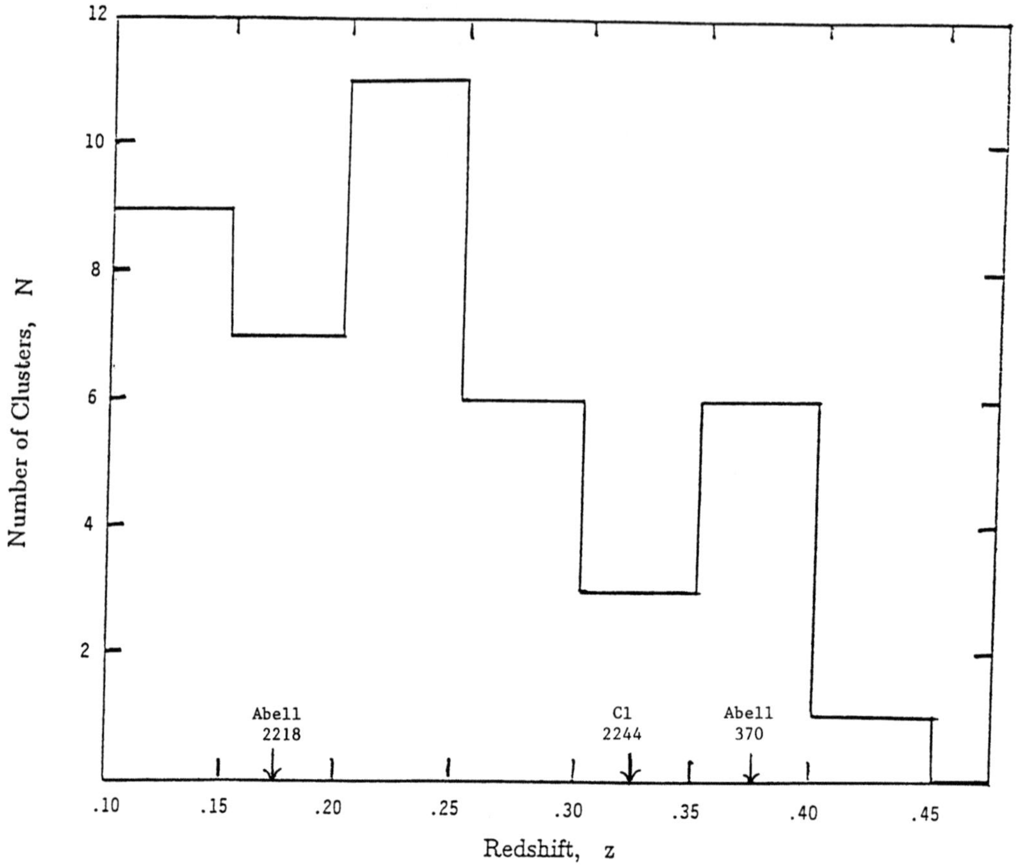

Fig. 1. Redshift distribution of clusters observed by video camera.

not observed before. Better data were needed to ascertain their spectral and morphological properties. Figures 2 and 3 show the results of observation in 1985 by Roger Lynds using various CCD's available at the KPNO 4-meter telescope. With this new data, we could quantify some of the properties of these features, in particular, their extent and almost perfect circular geometry. This led us to announce their discovery (Lynds and Petrosian 1986). The properties summarized in section II below are based on this recent data.

The filamentary feature in one of these clusters, Abell 370, has been observed by A. Hoag (1981), is evident in a picture published by Butcher, Oemler and Wells (1983), who did not comment on it, and is described as a ring around the southern cD galaxy by Soucail *et al.* (1987a) in the first of their publications on this object. All these authors were evidently unaware of our past and recent observations.

As we shall see, these observations already provided strong support for the gravitational lens hypothesis, but the conclusive evidence came from spectroscopic observations. With Roger Lynds, we attempted polarization and spectroscopic observation of the arcs in Abell 370 and Cl 2244 during the 1986 fall observing session at the KPNO 4-meter telescope. Because of bad weather, a very limited set of data was obtained, from which we could only set uninteresting upper limits on polarization and strength of emission lines in Cl 2244. A spectrum of the arc in Abell 370 was obtained by Soucail *et al.* (1987b) during the same observing season. This spectrum was of lower quality than the subsequent spectra described below, which shows a strong emission line plus some absorption features. In this earlier spectrum the emission line was tentatively (but mistakenly) identified as the 4000Å break, giving rise to a redshift for this arc of $z = 0.59$.

In the next observing season, three independent spectra were obtained of the arcs. The observation by Miller and Goodrich (1988) from Lick Observatory, even though inconclusive about the presence of any spectral lines, set a stringent upper limit of 4.4% on the polarization of Cl 2244. Observations by us (Lynds and Petrosian 1988) and Soucail *et al.* (1988) of the Abell 370 arc are in excellent agreement, showing the presence of a strong line at 6429A (identified as [OII] λ 3727) and three or four absorption features (which can be identified with well-known features), indicating a redshift of 0.725 for the arc. Our spectrum of Cl 2244, on the other hand, shows only a possible emission line, so that no definite value for the redshift can be given. However, if we assume that, like in Abell 370, this line is also due to [OII] λ 3727 line emission, the redshift of the arc turns out to be $z = 0.83$.

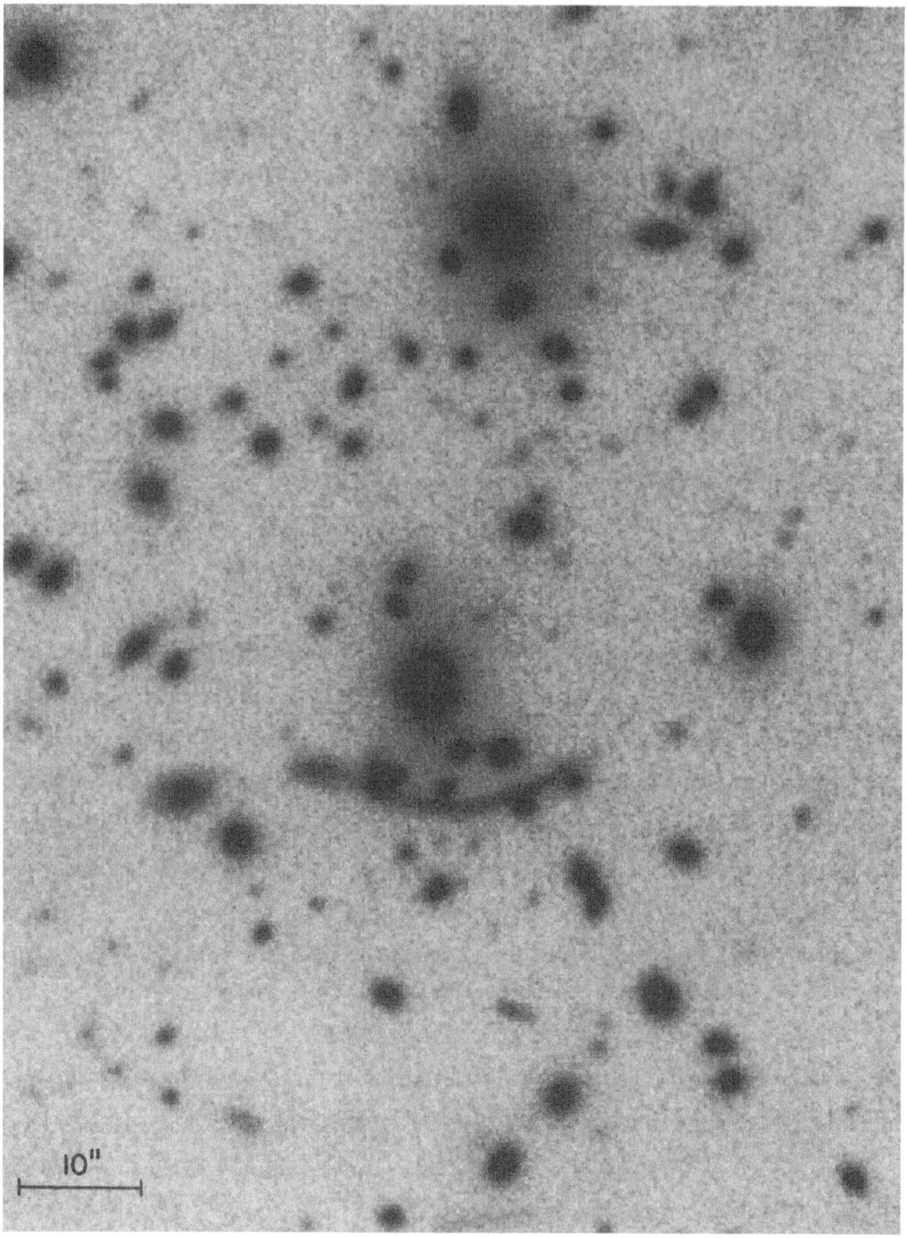

Fig. 2. CCD picture of Abell 370 from Petrosian and Lynds (1989).

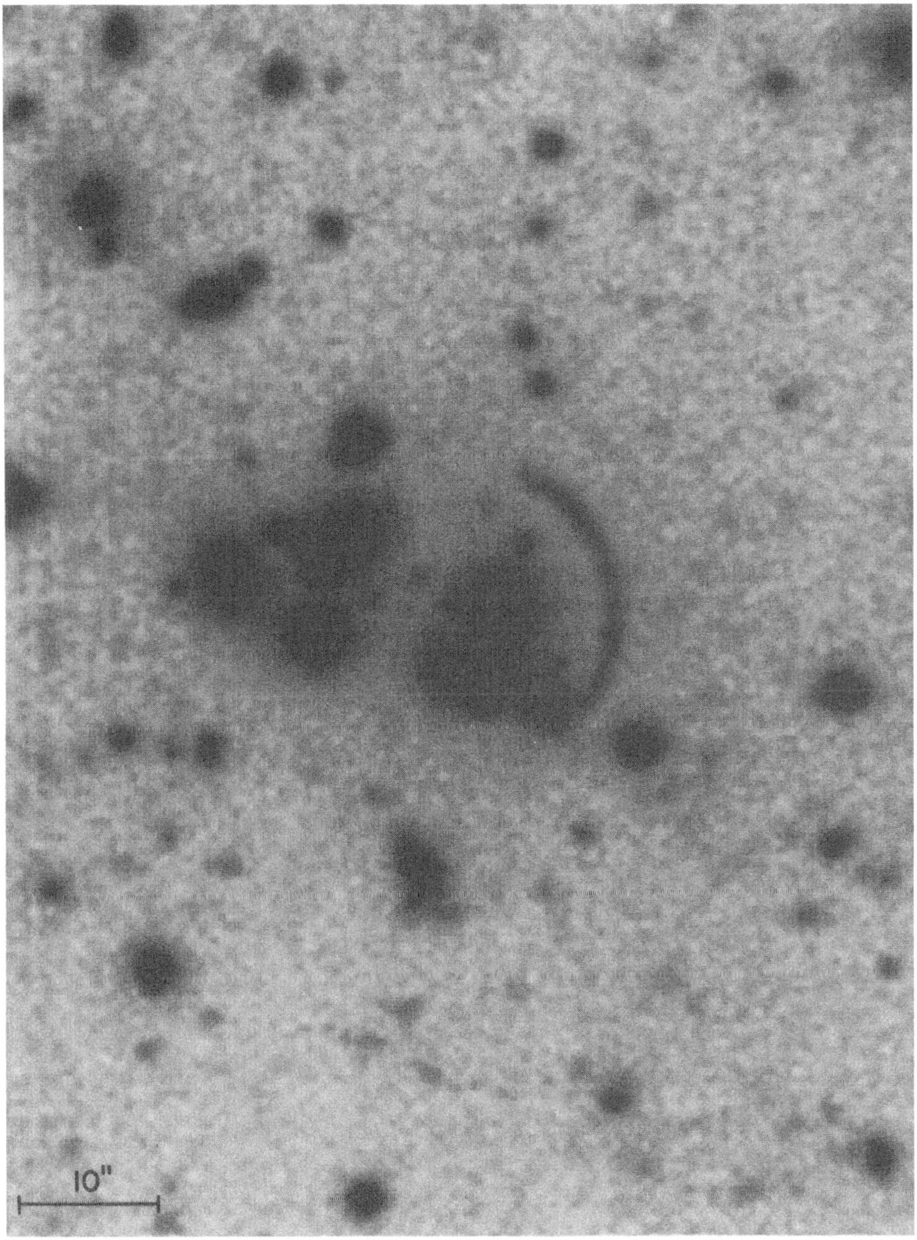

Fig. 3. CCD picture of Cl 2244 from Petrosian and Lynds (1989).

These higher redshifts for the arcs, as compared with the redshifts of the galaxies in the clusters where they are found, leave little doubt that the arcs are gravitational lens images of distinct sources produced by the cluster of the galaxies as envisioned by Zwicky (1937) half a century ago. The presence of similar filamentary features has been reported in two other clusters (Lavery and Henry 1988; D. Koo, private communications 1987). Undoubtedly many more, perhaps not as spectacular as those in Abell 370 and Cl 2244, will be discovered in the future.

Summary of Observations. The observed properties of the two arcs (Abell 370, Cl 2244) and their respective clusters are described in some detail by Lynds and Petrosian (1989). Some of this information will be reviewed here briefly. All the relevant observations are in the optical range. The observed X-rays from all three clusters are typical of emission from the hot extragalactic gas found in all rich clusters, and our VLA observations have shown a few interesting sources that appear unrelated to the arcs. The upper limit of few tens of μJy on the radio radiation from the arcs is an indication of the low radio emission from the source. Analysis of IRAS data by T. Soifer (private communication) has revealed only one interesting source near Abell 2218, which most likely is a foreground spiral galaxy.

The optical observations, however, contain a wealth of relevant information, only some of which will be described here and used in the next section for modeling purposes. The arcs are the largest coherent optical features observed and are extraordinary in their degree of circularity and essentially uniform width along their lengths. At the redshift of the clusters, the observed length of $\approx 20''$ and width of $\approx 1''$ implies a linear length of ~ 100 kpc (and a similar radius of curvature) and a width of 5 kpc. (Here and in what follows we assume the cosmological parameters $\Omega = 1$, $\Lambda = 0$, $H_o = 50$ km s^{-1} Mpc^{-1}.) The arc in Abell 370 is resolved in the direction perpendicular to its length and shows a factor of about 3 increase in intrinsic width from infrared frequencies to ultraviolet frequencies. The arc in Cl 2244 is essentially a segment of a perfect circle and shows some surface brightness variation along its length (the southern knot may not be part of the arc). A few galaxy images are superimposed on the Abell 370 arc, making it difficult to determine the existence of any surface brightness variations. But in this case there is clear evidence of deviation from circularity at both ends of the arc, especially in the eastern end where a prominent elongated knot deviates considerably southward from the circular arc.

The wideband spectra show the arcs to be, in general, much bluer than the galaxies, and Abell 370 has extremely large ultraviolet (rest wavelength $\simeq 2400$Å) emission. The galaxy spectra can be fitted to black body spectra of temperature 4500 K and the Abell 370 arc can be fitted with a power law spectrum of spectral index -2. The arc in Cl 2244 was not detected with the U filter. Figure 4 summarizes these results.

The visual magnitude of the arcs are $\simeq 20$, comparable to that of the cD galaxies. At the redshift of the clusters, these correspond to visual luminosities of about $10^{10}\ L_\odot$. The $B-V$, $V-R$ and $R-I$ colors would indicate a bolometric correction and K-correction of 1.5 mag and a mass-to-light ratio $M/L \simeq 4 \times M_\odot/L_\odot$ and therefore a visible stellar mass of $1.5 \times 10^{11}\ M_\odot$. At the higher redshifts of the arcs and with an assumed magnification of 20 due to gravitational lensing, the bolometric luminosity and mass of the sources are approximately $10^{10}\ L_\odot$. The observed mean surface brightness is 23.2 visual mag. per sq. arcsec. This corresponds to about 22 and 21 visual mag. per sqårcsec at the redshift of the clusters and of the arcs, respectively. The latter is considerably larger than the mean surface brightness of disk galaxies.

The high resolution spectra also indicate an unusual nature of the emission from the arcs and, as mentioned above, imply sources much farther away than the clusters. This clearly involves the assumption that the observed emission line in Cl 2244 is due to the [OII] λ 3727 line. Because of the weakness of the absorption features in the spectra of the Abell 370 arc and the fact that at the redshift of this cluster the observed emission line coincides with the HeII λ 4686 line position, some doubt has been raised about the validity of the redshift 0.725 for this arc. Under normal circumstances one would expect $H\beta$ to be as strong and the [OIII] λ 5007 line to be much stronger than the HeII line. For the [OIII] line to be absent, one would require an extremely low gas temperature, low abundance of oxygen, or some unusual ionization condition (Katz 1988). Assuming that such lines can be hidden, we estimate that the observed line strength of 10^{-15} erg cm^{-2} s^{-1} implies a total mass of $3 \times 10^{10}\ M_\odot$ for the gas (consisting of 10 percent He) within the region where the He is fully ionized if the arc is at the distance of the cluster. This is a large fraction (30%) of the stellar mass derived above. Combined with the fact that the colors and line to continuum strength ratios are nearly constant along the arc (including the knot at the eastern end), the hypothesis that the emission from the arc is due to an object local to the cluster seems artificial. We, however, believe in the reality of the absorption lines and the identification of the emission line with [OII] λ 3727, which at the redshift of 0.725 and magnification of factor 20 implies a more reasonable gas mass of $\sim 2 \times 10^8\ M_\odot$ for temperature $T = 10^4$ K and solar abundance of oxygen. In what follows we shall, therefore, assume redshifts of 0.725 and 0.83 for the arcs in Abell 370 and Cl 2244, respectively.

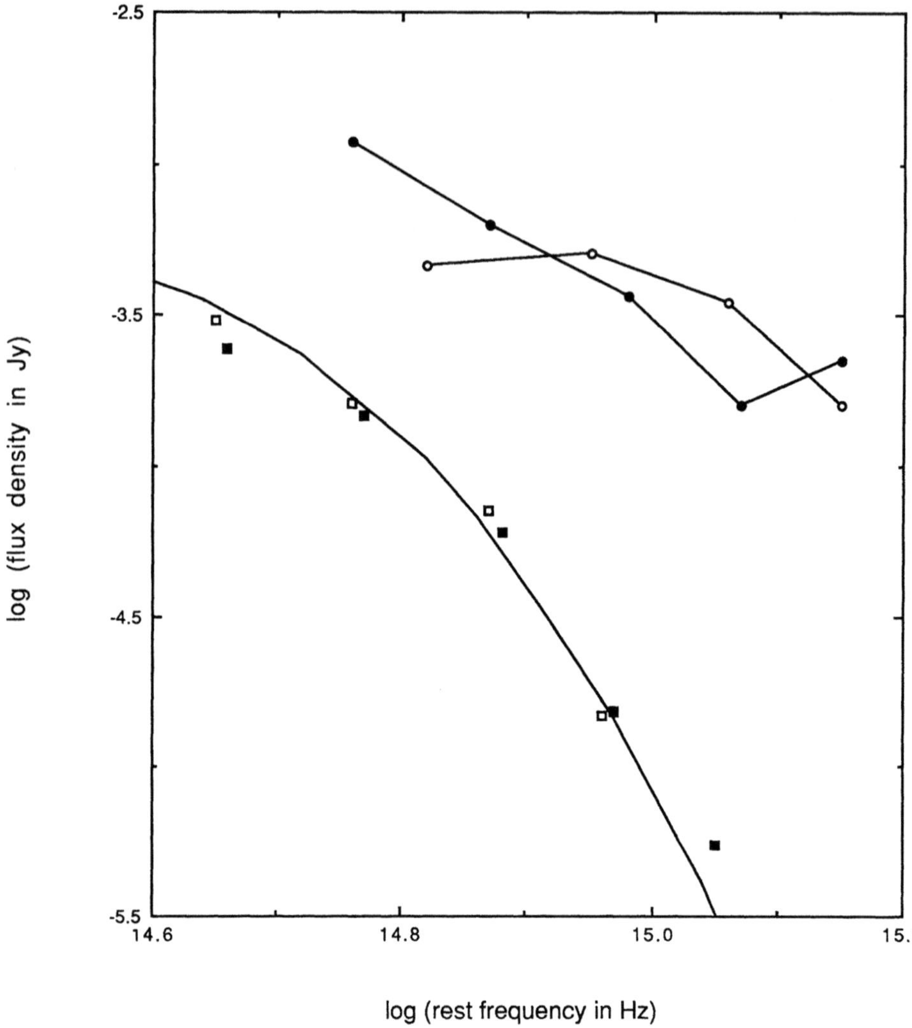

Fig. 4. The spectrum of cD galaxies (squares) and the arcs (circles) from wideband photometry. The frequency is that at rest with respect to the cluster or the source with redshift 0.725 for Abell 370 (filled symbols) and 0.83 for Cl 2244 (open symbols). The arc spectra are shifted upward by a factor of 10. The continuous line for the galaxies is for a black body of $T = 4500$ K.

The two clusters are unusual compared with the rest of the clusters in our original list in that they seem to have two centers of mass concentrations. This is obvious for Abell 370, which has two identical cD galaxies separated by about 34″. The cD galaxy in Cl 2244 is located near the center of the arc with a few satellites around it. There is a similar concentration of galaxies with comparable luminosity at a distance of 20″ to the east of the cD. This nonspherical aspect of the distribution is important in the modeling, which we turn to now.

Models of Gravitational Lens. As described by Lynds and Petrosian (1989), the above observations, especially the high luminosity, the large length, circularity, and the approximate constancy of the width and surface brightness along the length of the two arcs, are inconsistent with what one may call more conventional models, such as a nonthermal (curved) jet, condensation in shocks (due to explosive or less violent events), and accretion and condensation in the wake of a gravitationally orbiting object. The observed high redshift of the Abell 370 arc (and by inference that of Cl 2244) plus the low degree of polarization, observed by Miller and Goodrich for Cl 2244, render these models and the light echo model of Katz (1987) untenable, leaving the gravitational lens model as the most probable explanation of the phenomenon. Therefore, for the first time, we have seen a well-resolved portion of the so-called "Einstein Ring," which is obtained when an extended source is located almost directly behind a point mass or behind an extended mass with spherically symmetric distribution. The mass required is $M \simeq 2.8 \times 10^{14} M_{\odot}(\theta/20'')^2 f$, where θ is the apparent angular radius of the ring and f depends on the cosmological model and the redshifts of the source and the lens; $f = 1$ and 0.7 for Abell 370 and Cl 2244, respectively. Considering the luminosities of $\simeq 10^{11}\ L_{\odot}$ of the cD galaxies, we see that there clearly is a need for large quantities of dark matter giving rise to $M/L \simeq 10^3\ M_{\odot}/L_{\odot}$.

The observed arcs, however, do not form a complete ring, indicating a slight displacement of the lensing mass from the line of sight to the source. In this case, one would expect two approximately equal arcs. The absence of the second arc can be explained by a lens with nonspherical mass distribution such as the two point mass lens model investigated by Schneider and Weiss (1986) or by an extended lens with elliptical gravitational potential (Narayan and Blandford 1987). In fact, any deviation from spherical symmetry of the lens changes the degenerate point caustic of a spherical lens to a caustic with cusps such that multiple images merge into a single elongated arclike feature whenever the source covers a fair portion of a cusp.

As pointed out above, the clusters in both cases with well-defined arcs appear to have two centers of visible mass concentraton rather than the more common clusters domitated by one giant cD galaxy as the primary center of the visible mass. If the visible galaxies in the clusters trace the total mass, then in both clusters we have a situation akin to that explored by Schneider and Weiss (1986). In the case where the dark matter distribution does not follow that of visible galaxies, but consists of a spherically symmetric mass located near the center of curvature of the arc, the overall distribution of dark and visible matters will be sufficiently nonspherical to give rise to only one prominent arc. In what follows, both of these possibilities are explored, with the assumption that the background source is a simple circular disk with a uniform surface brightness. For an assumed set of parameters for a thin lens, the lens equation has been solved numerically by Anton Bergmann (Bergmann and Petrosian 1987; Bergmann, Petrosian, and Lynds 1988) using the ray shooting technique (see, e.g., Schneider 1984). In this manner, the source size and location and the lens parameters that reproduce the observed shape and size of the arcs are determined.

i) Models for Cl 2244-01. The bolometric luminosities (assuming a bolometric and K-correction of 2.2 magnitudes) of nine objects in the vicinity of the arc in this cluster are given in the second column of Table 1. The galaxies are numbered counterclockwise starting with the cD galaxy as shown in Figure 5a. The other galaxies in the field are either too small or too far from the arc to affect its shape. For simplicity, in some cases (e.g., objects #2 and 7), we have combined a few closely located objects as one mass concentration. The separation of these objects into their constituent galaxies of the same total mass will not alter the results presented below significantly.

As is evident with mass-to-light ratios of a few hundred (in solar units), believed to be the characteristics of some galaxies and clusters as a whole, the total mass of the nine objects is comparable to the required mass of $\sim 10^{14}\ M_\odot$ mentioned above. Figure 5b shows an image produced with the almost constant mass-to-light ratio of $M/L \simeq 220\ M_\odot/L_\odot$ except at positions 3 and 4. The masses and mass-to-light ratios are shown in columns 3 and 4 of Table 1. Of these objects, the masses at positions 1 and 2 are the primary lens and the masses at positions 7, 8 and 9 provide the requisite nonsphericity so that the second image located in the middle of these two centers of concentrations will be small and invisible or hidden. Masses at positions 5 and 6 produce minor modifications in the northern portion of the arc. The objects on positions 3 and 4, being very close to the arc, have significant effects. Consequently, smaller masses can be tolerated at these positions. The values given in Table 1 for these objects are the maximum allowed. These objects may be foreground sources with much lower luminosities and

Table 1

Model Parameters for Cl 2244 Lens

Galaxy	$\frac{L_{bol}}{10^{11}L_\odot}$	Model 1 $\frac{M}{10^{13}M_\odot}$	Model 1 $\frac{ML_\odot}{M_\odot L}$	Model 2 $\frac{M}{10^{13}M_\odot}$	Model 2 $\frac{ML_\odot}{M_\odot L}$	Model 3 $\frac{M}{10^{13}M_\odot}$	Model 3 $\frac{ML_\odot}{M_\odot L}$
1	0.9	2.0	220	3.4	380	0.20	20
2	1.2	2.5	210	0.77	64	0.26	20
3	1.2	<0.03	3	<0.04	3	<0.03	3
4	0.4	<0.03	8	<0.02	5	<0.015	4
5	0.03	0.06	200	0.005	20	.0065	20
6	0.5	0.51	100	0.25	50	0.11	20
7	0.9	2.0	220	0.73	80	0.21	20
8	1.0	2.1	210	0.79	80	0.22	20
9	1.0	2.1	210	0.61	60	0.22	20
ext. mass	0	0	0	0	0	5.5	∞

corresponding higher mass-to-light ratios. (Note that the scaling factor f on the mass estimate given above decreases with redshift, making the maximum allowable masses smaller if these are foreground objects. However, the luminosity decrease will be larger.)

This model, however, is not unique. Other sets of masses and source locations can also produce acceptable images. Figure 5c shows another example with corresponding masses and M/L's given in columns 5 and 6 of Table 1. In this case, we have assigned the maximum possible mass to position 1 and the minimum mass to positions 7, 8 and 9, which produce acceptable curvature and sufficiently small secondary images.

Finally, in Figure 5d we present the results from a third model (columns 7 and 8 in Table 1) where the primary lens is considered to be an extended region of dark matter centered at the center of curvature of the arc. The visible galaxies now with much smaller mass-to-light ratios break the degeneracy and render the second arc invisible. This model again requires the large overall mass-to-light ratio and concentration within a hundred of kpc. This limitation on the degree of concentration will be described more fully in the models for the arc in Abell 370, to which we turn now.

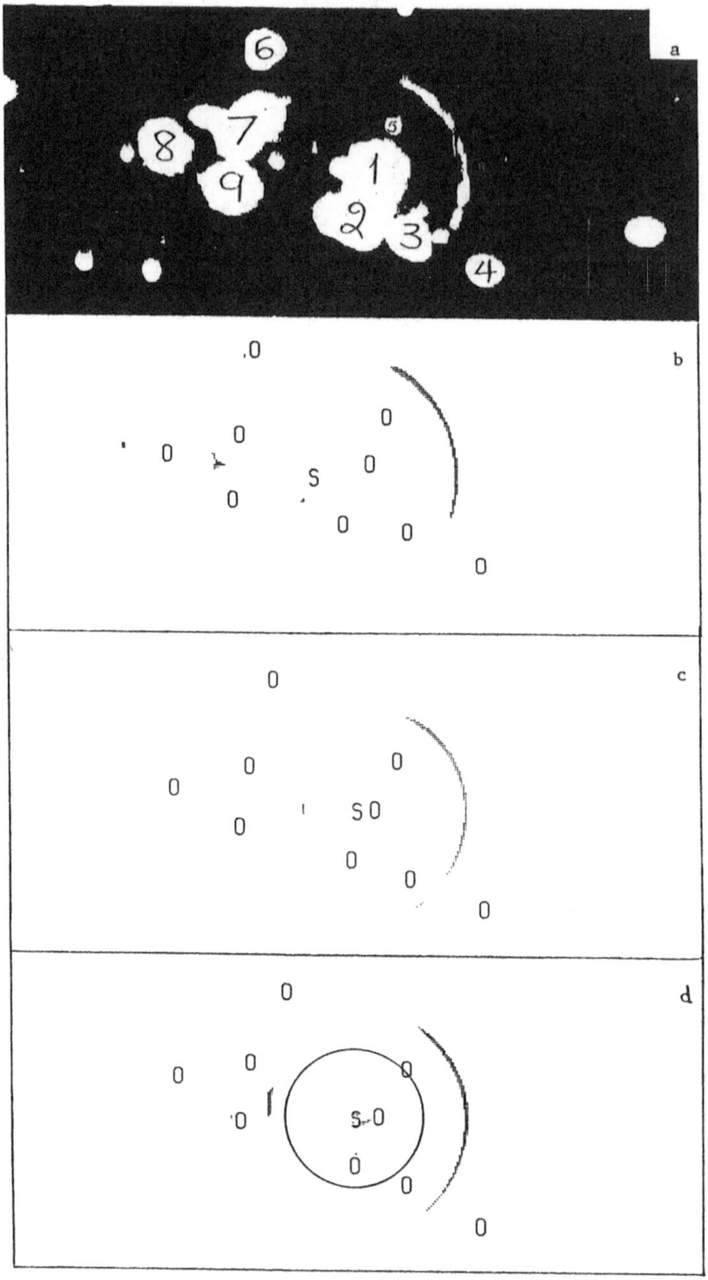

Fig. 5. Gravitational lens models for the arc in Cl 2244. See Table 1 for the value of parameters of the models. a) The arc and the relevant galaxies. b) Model 1 with constant mass-to-light ratio for all galaxies except 3 and 4. O = galaxy; S = source. c) Model 2 with variable mass-to-light ratio. d) Model 3 with a dominant dark component at the center of curvature of the arc almost directly in front of the source S. The circle represents the scale of the distribution r_o containing 70% of the total mass.

ii) Models for Abell 370. This cluster is dominated by two almost identical giant cD galaxies with visual magnitudes of 19.5 and bolometric luminosities $L = 10^{11}\ L_\odot$ (assuming bolometric and K-correction of 2.2 magnitudes). These galaxies (marked #1 and 2 in Figure 6a) and the dark matter will be the important masses in shaping the arc in this cluster. However, the smaller galaxies (#'s 3 to 7) near and superimposed on the arc are also important in this consideration. The fact that these galaxies do not distort the shape of the arc drastically can be used to set upper limits on the masses of these objects.

The models that assume that light is a good tracer of the total mass fail for this case. The primary reason for this failure is the larger separation of the two cD galaxies as compared with the radius of the curvature of the arc. Since the center of the arc is near galaxy #2, we need about $10^{14}\ M_\odot$ here. If galaxy #1 is also of the same mass, its critical line (Einstein ring) and that of galaxy #2 will be disjointed: each galaxy then will act as a separate lens and one expects a second arc located on the opposite side of galaxy #2. This image can be diminished only if galaxy #1 is sufficiently massive so that its critical line merges with that of galaxy #2. We find that this happens only for masses exceeding $10^{15}\ M_\odot$, which is unreasonable. As a result of the large mass of galaxy #1, another substantial image appears to the left of this galaxy. Furthermore, the encroaching galaxies (e.g., #3 and 7) must have masses less than $2 \times 10^{12}\ M_\odot$, implying a wide range of mass-to-light ratio for the galaxies in this cluster. Figure 6b shows an image for this case. The masses and M/L's for this and the two models to follow are given in Table 2.

Table 2

Model Parameters for Abell 370 Lens

Galaxy	$\frac{L_{bol}}{10^{11}L_\odot}$	Model 1 $\frac{M}{10^{13}M_\odot}$	Model 1 $\frac{ML_\odot}{M_\odot L}$	Model 2 $\frac{M}{10^{13}M_\odot}$	Model 2 $\frac{ML_\odot}{M_\odot L}$	Model 3 $\frac{M}{10^{13}M_\odot}$	Model 3 $\frac{ML_\odot}{M_\odot L}$
1	1.1	530	48000	0.60	55	0.37	34
2	1.1	59	5400	0.60	55	0.37	34
3	0.4	0.07	18	0.12	28	0.004	1
7	0.17	0.07	41	0.047	28	0.004	2
ext. mass	0	0	0	44.2	∞	370	∞

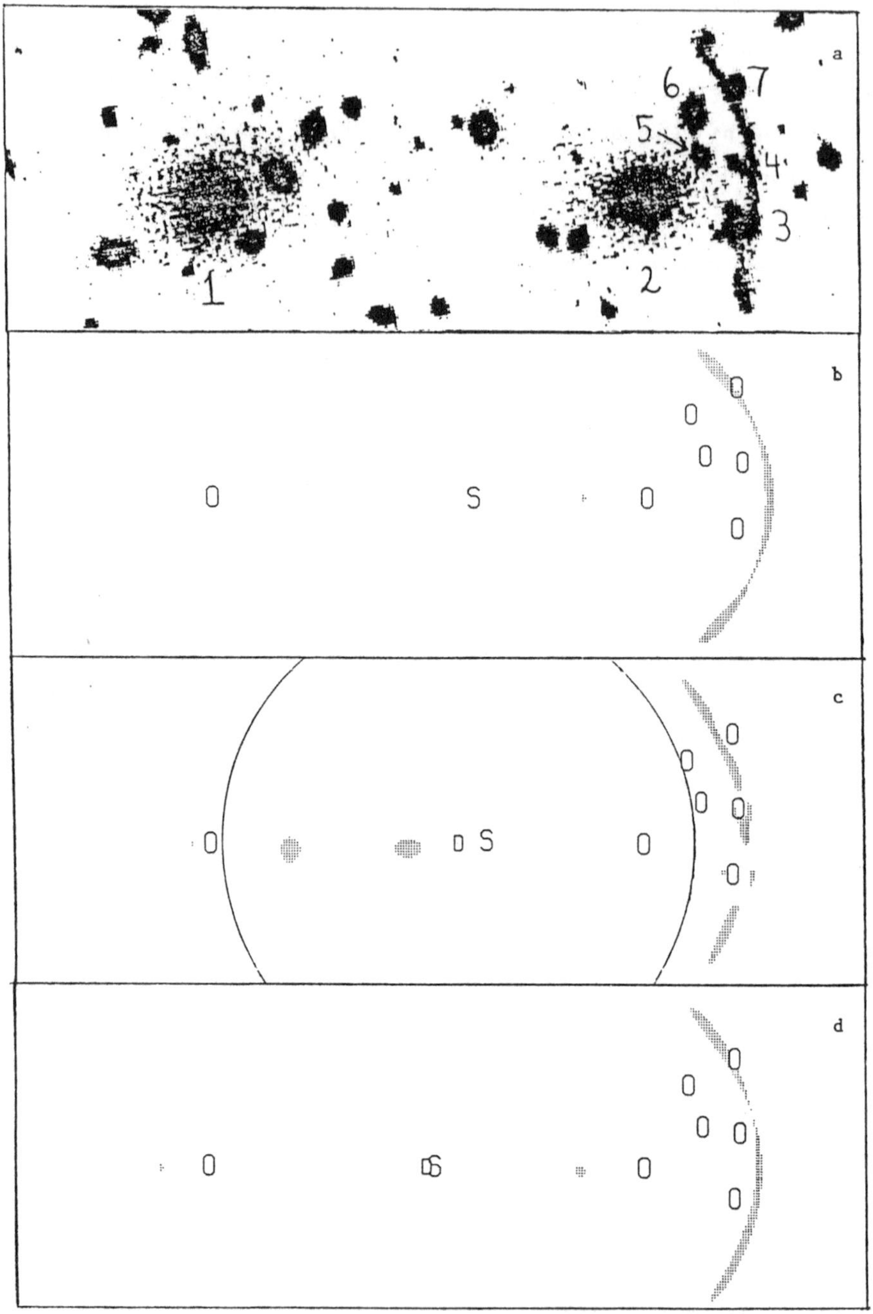

Fig. 6. Gravitational lens models for the arc in Abell 370. See Table 2 for the value of parameters of the models. a) The arc and the relevant galaxies. b) Model 1 consisting of two point masses. O = galaxy; S = source. c) Three-mass model with a dominant dark component D at the center of the curvature of the arc. The circle represents the scale of the distribution r_o containing 70% of the total mass. d) Same as c) with the dark component scale r_o almost four times larger.

A more reasonable set of parameters is obtained if the dark matter does not follow the light distribution but is concentrated near the center of curvature of the arc halfway between galaxies 1 and 2 in Figure 6a. This model is similar to one model considered by Soucail *et al.* (1987b). The dark matter now will be assumed to form an extended lens characterized by the central surface density Σ_o and scale r_o; $\Sigma(r) = \Sigma_o f(r/r_o)$. The total mass within a radius r is $M(r) = 2\pi r_o^2 \Sigma_o F(r/r_o)$, where $F(x) = \int_o^x f(x)x\,dx$. Figures 6c and 6d show two images produced by two different values for the parameters Σ_o and r_o. In the first of these, r_o is less than the radius of curvature, r_a, of the arc so that the dark matter can be treated essentially as a point mass. In the second one, $r_o > r_a$. It can be shown analytically and from numerical simulations (Bergmann, Petrosian, and Lynds 1988) that as r_o increases beyond r_a the extended dark component begins to approach a uniform distribution and required surface densities Σ_o or $\Sigma(r_a)$ approach the critical surface density $\Sigma_{cr} = (c^2/4\pi G)(D_s/D_d D_{ds})$, where D_s, D_d and D_{ds} are, respectively, the angular diameter distances to the source, lens, and the difference between these two distances. As a result, the magnification perpendicular to the arc becomes large, demanding a much smaller size for the source. This reduces the probability of alignment of the source and lens. The probability of finding arcs is further reduced in this case because only a limited number of combinations of source and lens redshifts can give rise to the equality $\Sigma_{cr} \approx \Sigma_o$ needed for production of arclike images. It, therefore, follows that most of the required dark matter must be located within the arc ($r_o < r_a$).

Summary and Conclusions. It has been shown that the arcs in two clusters of galaxies and pieces of elongated features in a third one are gravitational lens images. There are two important aspects that distinguish these cases of gravitational lensing from those involving quasars. First of all, the source being imaged is an extended (or resolved) galaxy-type object as compared with a pointlike quasar. This is probably also the case for the ring discovered by Hewitt *et al.* (1988) and discussed elsewhere in this proceeding. Secondly, the lens is a visible rich cluster rather than a barely detected (or presumed) field or cluster galaxy. Because of these differences, the phenomenon can be used more readily to study the mass distribution in the clusters and structure of the background sources, as demonstrated here.

The models described here have addressed the first of these two aspects and clearly show that dark matter is required at the centers of clusters of galaxies and that this dark matter either is associated with central galaxies with large mass-to-light ratios or is condensed at the very core of clusters. These conclusions are based on the shape and size of the arc. Fitting the more subtle features of the arcs, such as surface brightness fluctuation in Cl 2244 or deviation from the circular path in Abell 370, will require a refinement of the model, which will yield further insight on the mass distribution of clusters and especially the structure of the source. This analysis will be described in detail elsewhere (Bergmann, Petrosian, and Lynds 1988). Here, we comment briefly on these aspects.

Let us first consider the surface brightness fluctuation in Cl 2244. For a nonspherical lens, an arc is produced from merging of three images whenever the source covers a cusp. Thus, for a source that does not have a constant surface brightness, we expect some variation in the surface brightness of the arc. Similarly, variation in color or spectrum along the arc is also expected. If this is proven to be the case, then a consistent picture will yield detailed information about the source. In fact, once the details of the lens are derived from the general morphology of the arc, one can use the technique described by R. Blandford in these proceedings to invert the image and obtain the structure of the source. Another possible explanation for the surface brightness fluctuation is mini-lensing by low mass, low surface brightness objects lying along the ray path from the source to us. It can be shown that such small perturbations of the lens tend to distort the image along the arc but not perpendicular to it, so that the circular shape is maintained but gaps appear in the arc at the projected position of such mini-lenses. This, combined with effects of seeing, could give rise to observed fluctuations. The mass of a mini lens must be at least $10^{10}\ M_\odot$, in order to produce a detectable fluctuation, and its apparent surface brightness should be less than 23 visual magnitude per sq. arc sec. If the M/L for these objects is also as high as those in the rest of the cluster, their luminosities will be less than $10^8\ L_\odot$ so that they will not be detectable at the redshift of the cluster. A uniform distribution of such objects, as estimated by the degree of the fluctuation in the arc, could have a significant ($> 10^{14}\ M_\odot$) contribution to the total mass of the clusters. Of course, such mini-lenses do not have to be at the redshift of the cluster; they could have any redshift up to that of the source. In this case, a uniform distribution of them throughout the universe could make a significant contribution to the density parameter ($\Omega \geq 0.1$).

The deviation from circularity of Abell 370 will be more difficult to explain precisely because of the same property of gravitational lensing that led to the possibility of mini-lensing described above. It would be difficult to produce the shape of the arc near the eastern end with perturbing masses without breaking it into smaller individual images. Besides, there is no obvious candidate for such a perturbation. This deviation may, therefore, be an indication of a more complicated source structure than the simple circular disk assumed in the models described here.

Finally, current views on large scale structure of matter also appear to be in conflict with the rate of observations of arcs or elongated features. We have seen two well-defined arcs in two cases out of at most 30 objects. We estimate a probability of 0.1 for occurrence of such length arcs. The expected value for this probability depends on many factors, primary among which is the density of background sources. For sources as numerous as galaxies, this probability is expected to be less than 10^{-2}. A lower probability is obtained if one assumes that because of their unusual color, spectra, and relatively high surface brightness the sources may be rarer than galaxies. Furthermore, with a 10 percent chance of observing $20''$ long arcs, the chance of seeing shorter arcs is much higher. The arcs in Abell 2218, Abell 963, and pieces in Abell 370 described by G. Soucail in these proceedings (see Fort *et al.* 1988) may be in this category. Since images less than a few arc seconds long will, in general, be ignored, the absence of reports on smaller arcs perhaps is not surprising. As we know very little about the statistics of the galaxies at redshift $z > 0.5$, the above discrepancies may not be as serious as they seem and, in fact, the observation of these phenomena, when combined with deep surveys such as those by Tyson (1988), can be a good way to study the nature of high redshift galaxies.

Clearly, more observations and more theoretical study can reveal much more about the nature of high redshift sources and clusters of galaxies.

Acknowledgment. I would like to thank Roger Lynds, who obtained most of the data discussed here, and Anton Bergmann for his help in the theoretical modeling. This work was supported by NASA grants NCC 2-322 and NGR 05-020-668.

References.

Bergmann, A.G., and Petrosian, V. 1987, *Bull. AAS*, **19**, 1081.

Bergmann, A.G., Petrosian, V., and Lynds, R. 1988, (in preparation).

Butcher, H., Oemler, A., Jr., and Wells, D.C. 1983, *Ap. J. Suppl.*, **52**, 183.

Fort, B., Prieur, J.L., Mathez, G., Mellier, Y., and Soucail, G. *Astr. Ap.*, **200**, L5.

Hewitt, J.N., Turner, E.L., Schneider, D.P., Burke, B.F., Langston, G.I., and Lawrence, C.R. 1988, *Nature*, **333**, 537.

Hoag, A. 1981, *Bull. AAS*, **13**, 799.

Katz, J.I. 1987, *Astr. Ap.*, **182**, L19.

Katz, J.I. 1988, (preprint).

Lavery, R.J., and Henry, J.P. 1988, *Ap. J. (Letters)*, **329**, L21.

Lynds, R., and Petrosian, V. 1986, *Bull. AAS*, **18**, 1014.

Lynds, R., and Petrosian, V. 1988, *Bull. AAS*, **10**, 644.

Lynds, R., and Petrosian, V. 1989, *Ap. J.*, (in press).

Miller, J.S., and Goodrich, R.W. 1988, *Nature*, **331**, 685.

Narayan, R., and Blandford, R.D. 1987, *Ap. J.*, **310**, 568.

Petrosian, V. 1976, *Ap. J. (Letters)*, **209**, L5.

Petrosian, V., and Lynds, R. 1989, (in preparation).

Schneider, P. 1984, *Astr. Ap.*, **140**, 119

Schneider, P., and Weiss, A. 1986, *Astr. Ap.*, **164**, 237.

Soifer, T. 1988, (private communication).

Soucail, G., Fort, B., Mellier, Y., and Picat, J.P. 1987a, *Astr. Ap.*, **172**, L14.

Soucail, G., Mellier, Y., Fort, B., Hammer, F., and Mathez, G. 1987b, *Astr. Ap.*, **184**, L7.

Soucail, G., Mellier, Y., Fort, B., Mathez, G., and Cailloux, M. 1988, *Astr. Ap.*, **191**, L19.

Tyson, A.J. 1988, *A. J.*, (in press).

Zwicky, F. 1937, *Phys. Rev*, **51**, 290 and 679.

A GRAVITATIONAL TELESCOPE IN ABELL 370: INDEED IT WORKS!

G. Soucail
Observatoire Midi–Pyrénées
14 avenue E. Belin
31400 Toulouse, France

Recently, new observational results were obtained by a group from Toulouse Observatory (B. Fort, G. Mathez, Y. Mellier, and G. Soucail) on the cluster Abell 370. From optical CCD imaging and low-resolution spectroscopy, we were able to confirm the gravitational origin of the giant luminous arc lying in the center of A370. Moreover, reobserving deeply the center of the cluster, we have found several unexpected "arclike" structures that may be very distant, gravitationally lensed galaxies, demonstrating that the idea of using rich clusters of galaxies as gravitational telescopes is no longer a dream. A review of these data is presented, as well as a discussion of the future of such observations.

The Giant Arc in A370.

Presentation of A370. This cluster, at a redshift $z = 0.374$, is the most distant one in the Abell catalogue. An extensive spectrophotometric survey of its galaxies has been performed by the Toulouse group since 1985, and the main results were published by Mellier *et al.* (1988). It is very rich and dominated by two giant elliptical galaxies (referenced as #20 and #35 in Mellier *et al.* 1988). Its velocity dispersion, derived from 46 velocity measurements, is $\sigma \simeq 1700 \pm 170$ km s^{-1}, leading to a virial mass-to-light ratio of $M/L_R \simeq 96 \pm 10$ (assuming $H_0 = 50$ km s^{-1} Mpc^{-1}.

Near the center, and just south of galaxy #35, the giant arc extends for more than 2″ and is marginally resolved in width. This peculiar feature, codiscovered by Soucail *et al.* (1987) and Lynds and Petrosian (1986), has excited many astronomers during the last two years, before its nature was fully understood. Is it a coherent structure belonging to the cluster, more than 100 kpc long and less than 10 kpc wide, or an image of a distant object distorted by the gravitational field of the cluster? The answers to such questions had to come from more detailed observations, so a large amount of data was collected on this arc.

The difficulty of that work was due mainly to the faintness of the arc. Even with an integrated magnitude $R \simeq 19.3$, the surface brightness is $\mu_R = 23.2$ mag/square arcsec, corresponding to 10% of the sky brightness. The arc is very blue with a color index $B - R = 1.70$. A special comment has to be made concerning the eastern extremity of this arc, which is more enlarged than the main part of the structure and does depart from the main circular shape. Some people have claimed that it could be a background galaxy superimposed on the arc, but its surface brightness being similar to the central part, whatever the bandpass of the filter, is a strong indication against this hypothesis.

Spectroscopy of the giant arc. In any case, the clue to the nature of the arc had to come from its spectrum and from a redshift determination. The data presented here come from observations performed in October 1987 at the 3.6-m ESO telescope at La Silla (Chile). We used the EFOSC facility with the multiaperture spectroscopic system PUMA. A curved slit was punched, which precisely reproduced the shape of the arc. This method has several advantages that have to be stressed: first, the signal is integrated all along the arc, resulting in an increase of the signal-to-noise ratio of the spectrum. In addition, the spectrum can be extracted from each point of the arc, and we were able to confirm that the eastern end of the arc has a similar spectrum to that of the central part and belongs to the same structure. Moreover, the spectrum of each of the galaxies superimposed on the arc was extracted, and the redshift measurement confirmed that they all belong to the cluster.

The spectrum of the arc was measured between 4000 and 7100Å, with a resolution of 200Å during six hours of integration time (Soucail *et al.* 1988). An analysis of this spectrum leads to the identification of a strong and narrow emission line at 6427Å, which corresponds to the [OII] λ3727 emission line at a redshift $z = 0.724$. Moreover, several absorption lines confirm this identification, namely, MgII λ2800, the CaII $\lambda\lambda$3933,3968 lines, the CN band at 3883Å and some Balmer lines, especially a significant Hδ line (see Fig. 1). This spectrum is typical of a late-type galaxy with a large amount of star formation in it, but without any sign of nuclear activity.

The main result of this observation is that it strongly supports the lensing hypothesis, and the arc can be considered as an exceptional case of gravitational lensing for at least three reasons. It is the first case of an observation of a large extension of an "Einstein ring." Since its discovery, several other cases were announced, most notably, the radio ring in MG1131+0456 (Hewitt *et al.* 1988). Furthermore, it is the first well-confirmed case of lensing where the source is a galaxy, i.e., an extended object. Finally, in order to intepret its large radius of curvature,

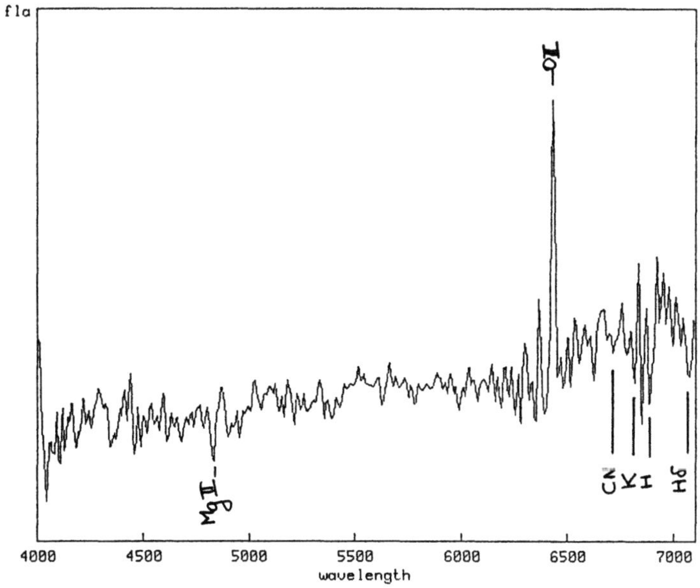

Fig. 1. Integrated spectrum of the giant arc.

one must invoke the whole cluster as the main deflector of the light, and this is the first case of lensing by a cluster of galaxies itself.

Detection of New Faint Distorted Structures in the Core of A370.

Observations. Several high resolution and deep CCD frames of the center of the cluster were obtained at the Prime Focus of the CFH Telescope in October 1987 by J. L. Prieur. The pixel size is 0.205″/pixel, well matched to a seeing of 0.7″ (FWHM) recorded during the observations. All the frames were added together in order to improve the detection of faint objects in the core of A370, and the B and R frames were calibrated for photometry. The first goal of these observations was to search for blue structures related to the giant arc as possible secondary images of the same source, mainly under the envelope of the giant galaxy #20. So to subtract all the galaxies of the cluster having a red color index, a $B - R$ frame was computed, where only the features significantly bluer than galaxy #20 remained (see Fig. 2). Three results can be extracted from this operation (Fort *et al.* 1988):

1. No "counter-arc" of any sort is detected near galaxy #20 or under its envelope, down to the detection limit.

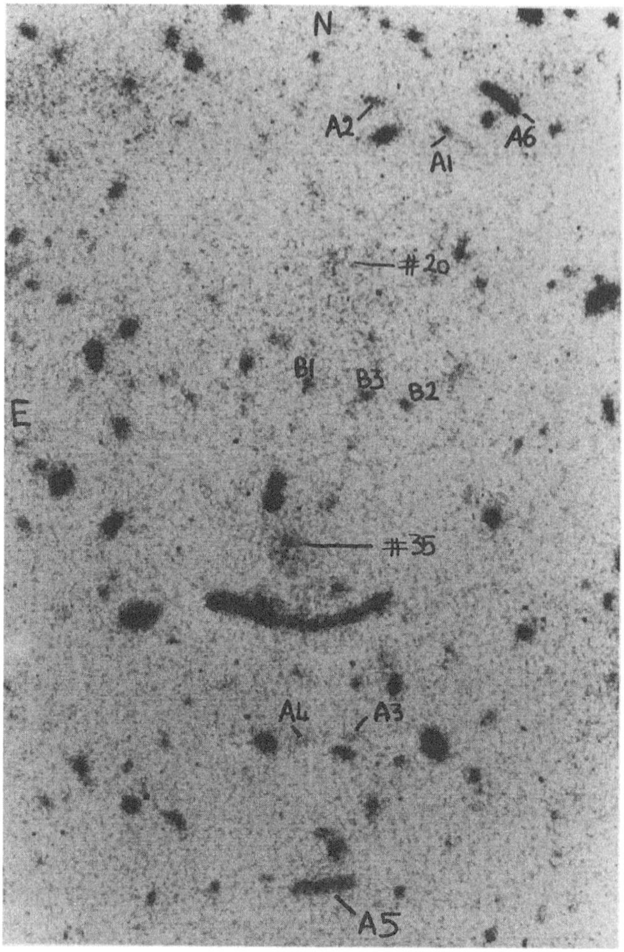

Fig. 2. Residuals of the subtraction of the R from the B frame. Most galaxies vanish, and several blue faint and elongated objects appear (A1 to A6) as well as three unresolved objects very near the center (B1 to B3).

2. Several "arclike" features are seen around the cluster core, located in concentric circles. Some of them (A1–A2 and A3–A4) seem to be linked into pairs, while A5 is the most elongated and distant one and A6 lies through a galaxy of the cluster.

3. Three "pointlike" objects, denoted B1, B2, and B3, having the same color index as the small arcs, lie very near the center of the cluster.

The main characteristics of these objects are the following: low surface brightness ($\mu_B \simeq 25$ mag/square arcsec) and very blue color indices: $B - R = 1.0$ for all of them, considering the uncertainties in the measurement, which reaches 0.3 mag in this magnitude range. They are

ID	B	B–R	μ_B	Length	Radius
A1	24.50	1.00	24.7	3"	48"
A2	24.05	1.05	25.0	4"	48"
A3	25.10	1.10	25.6	3"	37"
A4	25.05	1.00	25.2	2.5"	37"
A5	22.50	0.85	23.2	9.3"	56"
A6	23.30	1.00	24.0	5"	56"
Arc	22.40	1.70	23.6	22"	23"
B1	24.35	0.95	24.3	–	–
B2	24.35	1.15	24.6	–	–
B3	24.35	1.15	24.3	–	–

summarized in the Table, which indicates the identification, the B magnitude, the $B - R$ color index, the mean B surface brightness, the elongation, and the radius of curvature of these objects.

Discussion: Are these objects gravitationally lensed galaxies? After the confirmation of a gravitational effect for the giant arc, it is tempting to interpret the other structures in the same way. Several arguments strongly favor such a hypothesis.

First, considering their color and surface brightness, and comparing them with evolutionary models for galaxies, one can argue that the light should come from galaxies at redshift between 0.6 and 1.3, and most likely at $z \sim 1$ (see Fig. 3 in Fort *et al.* 1988). Such a view is strongly reinforced by the recent discovery by Tyson and his collaborators of a large population of extremely faint and blue objects in blank fields near the South Galactic Pole (Tyson 1988). They interpret this population to be galaxies with redshift between 1 and 2, having an important fraction of star formation in it. From their galaxy count, we can expect to see three galaxies with $B \leq 26$ in 300 square arcsec behind the center of A370. This number is quite consistent with the number of structures observed, especially if we consider some of them as multiple images of the same source (see, for example, A1–A2 or A3–A4).

In addition, Grossman and Narayan (1988) have presented numerical simulations of lensing of several extended sources by a cluster of galaxies. The distortion of the sources lying near the axis of the center of the cluster is strikingly similar to the shape of the structures detected in A370!

Of course, these arguments are not a strong confirmation of the gravitational lens hypothesis, but we must keep in mind that a redshift measurement is presently not possible for such faint structures, except for the "brightest one," A5. Indeed we plan to try to obtain its spectrum as soon as possible.

Moreover, if these new arcs are really gravitationally lensed galaxies, more can be said about the deflecting potential. First, since each of the structures is centered both on the optical center of the cluster and on the X-ray center, most of the deflecting mass probably lies inside the cluster and is not spread all along the line of sight. Then, identifying their radius with the critical radius, one can evaluate the mass necessary to create such a deflection. Except for A5, whose radius is too large, the mass-to-light ratio obtained from the other ones is again consistent with the virial M/L ratio of the cluster or the M/L ratio derived from the giant arc, reinforcing the idea that a large fraction of dark matter lies inside the clusters (Fort *et al.* 1988).

Conclusion. Since the first announcement of the discovery of the arcs, several new ones have been observed in several distant clusters: Cl2244-02 (Lynds and Petrosian 1986), A2218 (Pello-Descayre *et al.* 1988; Petrosian 1987, private communication), and A963 (Lavery and Henry 1988), among others. If indeed all of these cases are due to gravitational lensing, we can argue that lensing by extended sources is becoming a new field of research. As was predicted several years ago, the use of clusters of galaxies as "gravitational telescopes" is something that should be intensively exploited, as we have shown that it works. A long-term program of such observations is now in preparation, which is complementary to the one performed by Tyson. At least two consequences can be expected. Considering the study of the gravitational potential of the clusters, one can expect to derive some mass determinations (for the whole cluster, or for some individual galaxies), reinforcing the evidence of dark matter inside the clusters. The problem of its localization still remains but could be more constrained. On the other hand, the use of the gravitational magnification by the clusters will probably make possible observations of the distant universe, both in photometry (detection of highly redshifted galaxies) and in spectroscopy for the brightest ones, leading to the possibility of obtaining better spectra of very distant galaxies.

References.

Fort, B., Prieur, J.L., Mathez, G., Mellier, Y., and Soucail, G. 1988, *Astr. Ap.*, **200**, L5.

Grossman, S.A., and Narayan, R. 1988, *Ap. J. (Letters)*, **324**, L37.

Hewitt, J.N., Turner, E.L., Schneider, D.P., Burke, B.F., Langston, G.I., and Lawrence, C.R. 1988, *Nature*, **333**, 537.

Lavery, R.J., and Henry, J.P. 1988, *Ap. J. (Letters)*, **329**, L17.

Lynds, R., and Petrosian, V. 1986, *Bull. AAS*, **18**, 1018.

Mellier, Y., Soucail, G., Fort, B., and Mathez, G. 1988, *Astr. Ap.*, **199**, 13.

Pello-Descayre, R., Soucail, G., Sanahuja, B., Mathez, G., and Ojero, E. 1988, *Astr. Ap.*, **190**, L11.

Soucail, G., Fort, B., Mellier, Y., and Picat, J.P. 1987, *Astr. Ap.*, **172**, L14.

Soucail, G., Mellier, Y., Fort, B., Mathez, G., and Cailloux, M. 1988, *Astr. Ap.*, **191**, L19.

Tyson, A.J. 1988, *A. J.*, **96**, 1.

OBSERVATIONS OF THE BLUE ARCS IN ABELL 963

Russell J. Lavery
Institute for Astronomy, University of Hawaii
2680 Woodlawn Drive, Honolulu, HI 96822

Abstract. Initial spectroscopic observations of the two blue arcs in Abell 963 show no evidence for strong emission lines, though continuum radiation has been detected. Since these arcs are likely the result of gravitational lensing, this observation puts several constraints on the nature of the lensed object. The lack of any broad emission lines limits the possibility of the lensed object being a QSO unless it is at a redshift of greater than 5. If the lensed object is a galaxy with [O II] $\lambda3727$ emission, as might be expected from the arc color, the galaxy must be at a redshift of greater than 0.9. The lack of Lyman α emission constrains the redshift to be less than 2.5.

Introduction. The discovery of giant luminous arcs in the two distant clusters of galaxies, Abell 370 and Cl 2244-02 (Soucail *et al.* 1987; Lynds and Petrosian 1986) was greeted with much excitement in the astronomical community. Since their discovery, a number of different and interesting suggestions have been proposed for their origin. But, it now appears clear that the arcs are the result of gravitational lensing. Spectroscopic observations of the arc in A370 now give strong support to the the idea that the light of the arc comes from a galaxy at a redshift of 0.724 (Soucail *et al.* 1988; Lynds and Petrosian 1988) whose light path has been distorted by the core of the cluster. Additionally, Miller and Goodrich (1988) reached a similar conclusion by a process of elimination based on their imaging polarimetric observations of the arc in Cl 2244-02. What remained somewhat of a puzzle with the gravitational lens interpretation was that there appeared to be only a single image produced when two images were expected. But, now this difficulty has also been explained. Each of these clusters appears to have two central concentrations and therefore the simple assumption of spherical symmetry applied in most lens models is probably incorrect for these clusters. Modeling of asymmetric lenses by Grossman and Narayan (1988) has shown that such lenses can produce a single large arclike image.

In support of the gravitational lensing hypothesis, Lavery and Henry (1988) reported on the first observations of two blue arcs in the rich cluster of galaxies Abell 963. The geometry of the two arcs in this cluster, dominated by a single large cD galaxy, is that expected from a simple symmetric gravitational lens with the background object slightly offset from the line of sight (Refsdal 1964; Leibes 1964). From their photometry, Lavery and Henry found that each arc was extremely blue in color, which suggested the arcs were not actually members of the cluster. This extremely blue color, $B - R \sim 0.85$ in the Johnson system, is consistent with that of an irregular galaxy at a redshift of 1.

Here I present spectroscopic observations of the larger southern arc in A963. Though no redshift is obtained, these observations place constraints on both the nature of the lensed source and its probable redshift.

Observations. The spectrum was obtained on 1988 April 9 at the University of Hawaii 2.2 Meter Telescope of the Mauna Kea Observatory. The detector was the UV-flooded IfA/NSF TI 800^2 3-phase CCD (Hlivak, Henry and Pilcher 1984) used with a low resolution grism spectrograph. The exposure time was two hours. The slit was positioned in a north-south alignment across the cD galaxy and the arc. This was not the optimum orientation for obtaining a spectrum of the southern arc, but it did provide the redshifts of several galaxies in the cD halo. These redshifts will be important for future modeling of this lens.

The arc spectrum was extracted from the spectral image, as was a sky spectrum, which was then subtracted. Because the arc is inside the cD envelope, there is some contamination from the cD galaxy in the red end of the arc spectrum. For this reason, I have not attempted to measure the arc colors from this spectrum. But, this spectrum is sufficient for the detection of emission lines, if present. The spectrum has been linearized and corrected for atmospheric absorption and the spectrograph system response.

I have also tried to extract a spectrum for the smaller northern arc. But, since this arc is well inside the cD galaxy halo, the contaminating cD light dominates the light from the arc itself. It is extremely difficult to obtain a spectrum of this arc due to its proximity to the cD galaxy.

Data. The resultant arc spectrum is presented in Figure 1. It is obvious that continuum light is present, but there are no emission lines. This lack of emission lines makes a redshift determination impossible, but it does allow several constraints to be placed on the nature of the arc. If the arc is produced by gravitational lensing, the lack of any broad emission lines eliminates the possibility of the background object being a QSO unless the redshift is greater

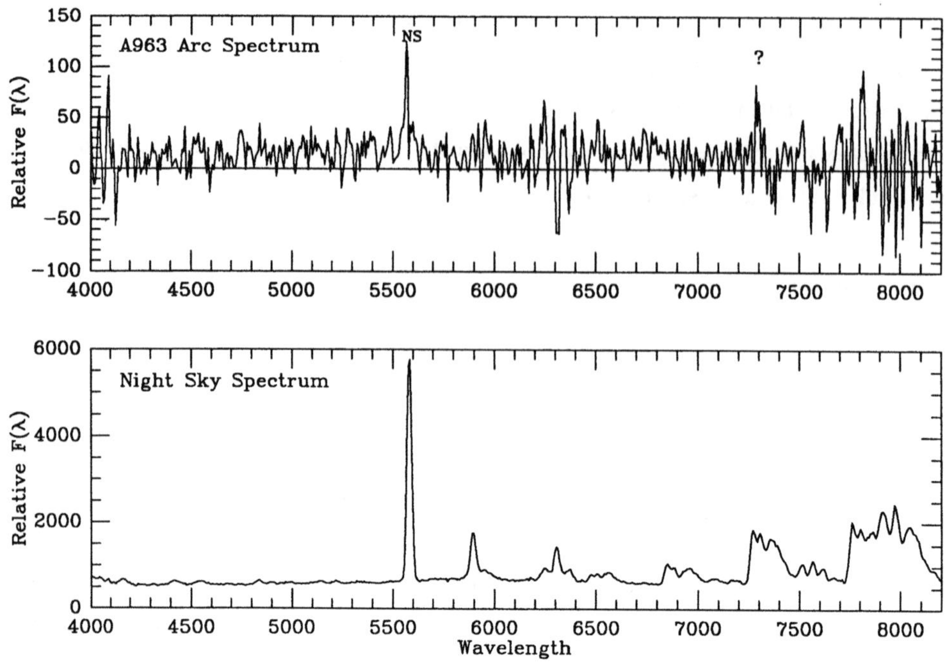

Fig. 1. Spectrum of the southern blue arc in Abell 963 and a night sky spectrum. Continuum radiation has been detected from the arc, but the spectrum lacks emission lines. The only possible emission line is identified by the ? at λ7296. Wavelength is given in Angstroms.

than 5. Table 1 lists the four strongest emission lines present in QSO spectra and the redshift range covered by each line in the wavelength region 4300Å to 7200Å. If any of these lines were present, they would have been detected. Of course, if the redshift were less than 0.54, other lines such as Hβ, [O II] or [O III] should have been detected. This suggests the lensed object is a background galaxy.

Table 1

Redshift Range for QSO Emission Lines

Line	Wavelength	Redshift Range
Mg II	2798	0.54 to 1.57
C III]	1909	1.25 to 2.77
C IV	1549	1.77 to 3.65
Ly α	1216	2.54 to 4.92

If the lensed object is a "normal" galaxy, it is probably at a redshift greater than 0.9. This result comes from both the arc color (Lavery and Henry 1988) and from the lack of an emission line in the spectrum. Any "normal" galaxy having such a blue color would undoubtedly have [O II] λ3727 present. Since no line is present, the line must be shifted beyond the observed wavelengths.

The only hint of a line in the spectrum is at λ7296. This is a very weak feature, but it is not the result of improper sky subtraction at the edge of the OH emission band at λ7251. If this line is real and assumed to be [O II], this would give a redshift of 0.957 for he lensed galaxy.

Discussion. The lack of emission lines in the spectrum of the arc is not all that surprising. Both the spectroscopic observations of the arc in A370 (Soucail *et al.* 1988; Lynds and Petrosian 1988) and the patchiness of the arcs in A963 (Lavery and Henry 1988) suggest that the arcs are the result of gravitational lensing of a background galaxy by these rich clusters of galaxies. Shortward of [O II] λ3727, normal galactic spectra are devoid of emission lines, probably out to Lyman α. This makes it difficult to obtain redshifts for galaxies at redshifts of 1 or greater, especially since the UV portion of galactic spectra is very uncertain and quite varied. From the color of the arc, Lavery and Henry (1988) determined that the background galaxy would be of redshift $\sim$ 1 or greater and this has been borne out by these spectroscopic observations.

As mentioned in the Data section, there is possibly a weak emission line at the wavelength of 7296Å. If this line is [O II], it would indicate a redshift of 0.957. But, the detection of this line is so unreliable that all one can really say is that it provides some optimism for future spectroscopic observations in determining the redshift of the arcs.

Because of the recent observations of a number of small "arcs" in the cluster A2218 (Pello-Descayre *et al.* 1988) and in A370 (see G. Soucail, these proceedings), I have looked for such structures in the images of A963 (see Fig. 2). There do not appear to be any obvious smaller "arcs" in this cluster. There is one possible feature several arcseconds to the south of the main arc, but it lies on a galaxy and it is, therefore, impossible to reach any definite conclusions about it. This lack of smaller "arcs" may be considered surprising, as one should expect there to be many background galaxies that would be lensed and that these smaller "arcs" in A2218 and A370 are the rule and not the exception. This question will only be answered through a deeper understanding of gravitational lensing by clusters of galaxies and through deeper imaging of Abell 963.

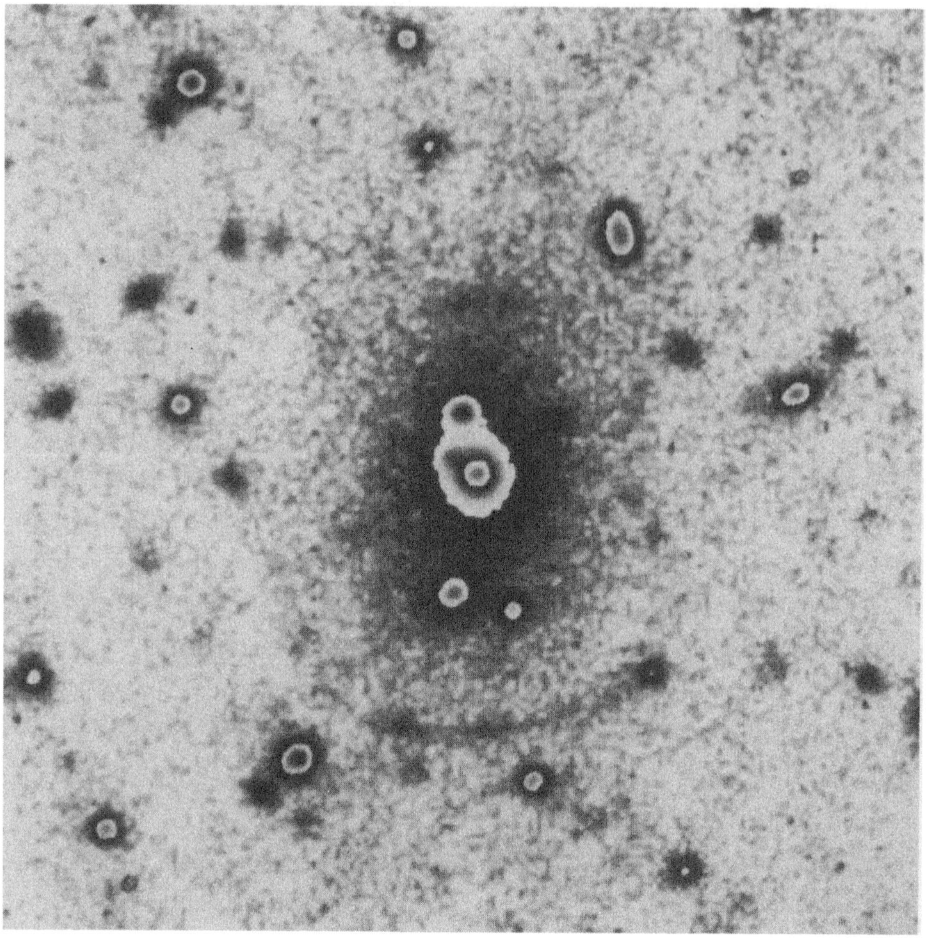

Fig. 2. B image of the two blue arcs in Abell 963. The separation of the larger southern arc from the cD nucleus is 18 arcseconds. North is up and east is to the left.

Conclusion. Spectroscopic observations of the blue arcs in Abell 963 show a lack of emission lines in the spectrum, though continuum light has been detected. This result eliminates the possibility of the lensed background object being a QSO. If the arcs are produced by the gravitational lensing of a background galaxy, it must lie in the approximate redshift range of 1 to 2.5.

References.

Grossman, S. A., and Narayan, R. 1988, *Ap. J. (Letters)*, **324**, L37.

Hlivak, R. J., Henry, J. P., and Pilcher, C. B. 1984, *Proc. Soc. Photo-Opt. Instr. Eng.*, **445**, 122.

Lavery, R. J., and Henry, J. P. 1988, *Ap. J. (Letters)*, **329**, L21.

Leibes, S. 1964, *Phys. Rev. B* **133**, 835.

Lynds, R. and Petrosian, V. 1986, *Bull. AAS*, **18**, 1014.

Lynds, R. and Petrosian, V. 1988, preprint.

Miller, J. S., and Goodrich, R. W. 1988, *Nature*, **331**, 685.

Pello-Descayre, R., Soucail, G., Sanahuja, B., Mathez, G., and Ojero, E. 1988, *Astr. Ap. (Letters)*, **190**, L11.

Refsdal, S. 1964. *M. N. R. A. S.*, **128**, 295.

Soucail, G., Fort, B., Mellier, Y., and Picat, J. P. 1987, *Astr. Ap. (Letters)*, **172**, L14.

Soucail, G., Mellier, Y., Fort, B., Mathez, G., and Cailloux, M. 1988, *Astr. Ap. (Letters)*, **191**, L19.

IS THE GIANT LUMINOUS ARC DUE TO LENSING BY A COSMIC STRING?

Xiangping Wu
Beijing Astronomical Observatory
Chinese Academy of Sciences
Beijing, 100080, China

Introduction. Three giant luminous arcs were discovered in three clusters of galaxies (Lynds and Petrosian 1986; Soucail *et al.* 1987*a*; Lavery and Henry 1988). There have been at least four conjectures proposed so far to explain these unusual features. The recent observations of the spectra of the arcs made by Miller and Goodrich (1988), however, have eliminated all but the gravitational lens hypothesis, which was first proposed by Paczynski (1987). He interpreted these features as being due to the gravitational lensing of background galaxies by the foreground clusters. The detection of a redshift of 0.724 in the arc associated with Abell 370 ($Z = 0.373$) (Soucail *et al.* 1987*b*) strongly supported this explanation. But the more detailed computer simulation made by Grossman and Narayan (1988) shows that (1) an unreasonably small core radius of the cluster is needed, and (2) several medium-sized arcs should be observed for each arc discovered. Therefore, problems still remain with this model. In this paper, I propose that the giant luminous arc might be due to the lensing by a cosmic string and present the computations based on the model in which the background galaxy acts as the source and the long-lived loop cosmic string as the lensing object.

Calculations and Results. Though the gravitational lensing properties of a straight cosmic string have been fully discussed by Gott (1985), no exact solution has so far been found for a closed loop string. In this work, the case of a light-ray perpendicular to the plane of the loop is considered. Using the linearized Einstein theory for the loop string, the bending angle of light with the impact parameter b will be ($c = G = 1$) $\alpha(b < a) = 0$ and $\alpha(b > a) = 8\pi\mu a/b$, where μ is the mass per unit length, and a is the ring radius. For the case of $b > a$ (outside the ring), the lensing properties of a ring are the same as those of a point mass. From the lensing equation, we can have $b_{\pm} = [\sqrt{\ell^2 + 4D} \pm \ell]/2$, where ℓ is the offset of the undeflected ray from the source at the deflector, $D = 8\pi\mu a D_d D_{ds}/D_s \equiv 8\pi\mu a d(Z_d, Z_s)/H_0$, D_d and D_s are the angular diameter distances to the deflection and source, respectively, D_{ds} is the difference between these two distances, and H_0 is the Hubble constant. If $b_{\pm} > a$, (1) two images will occur when $\ell < -a(1 - D/a^2)$, and (2) a single image will occur when $\ell > a(1 - D/a^2)$. For the

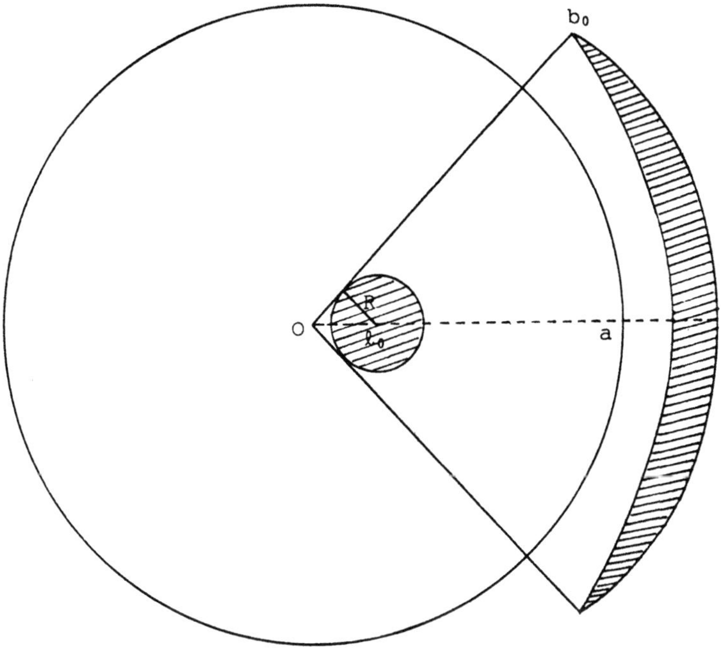

Figure 1. The arc-shape image of the circular disk of radius R with constant surface brightness that will be produced by a long-lived loop lens.

largest loop ($a \sim 2H_0^{-1}$), we can have $D/a^2 \sim 4\pi\mu d(Z_d, Z_s) \ll 1$, and only one image appears. For the smallest loop ($a \sim \mu H_0^{-1}$) and not very large Z_s, $D/a^2 \sim 8\pi d(Z_d, Z_s) < 1$. In this case, both single and double images are possible.

The remarkable feature of the loop lensing is that the images of the background source always lie outside the ring. So, the arc-shape radially oriented elongated image will occur for a source with radius R near the center of the ring (see Fig. 1). The arc length is $L = 2b_0 \sin^{-1}(R/\ell_0)$, where $b_0 = [\sqrt{\ell_0^2 - R^2} + \sqrt{\ell_0^2 - R^2 + 4D}]/2$ and can be taken as the mean radius of arc. From the observation of the arc in Abell 370, we have $b_0 \simeq 50$ kpc and $L \simeq 150$ kpc ($H_0 = 50$ km s^{-1} Mpc^{-1}). Thus, we have $\sin^{-1}(R/\ell_0) = L/2b_0 \simeq 1.5$, $R \simeq \ell_0$, and $b_0 \simeq \sqrt{D}$, respectively. Since the observations did not show the secondary image, $\ell > -a(1 - D/a^2)$. On the other hand, it is difficult to form a giant arc if $\ell > a$; thus, $D < 2a^2$. Therefore, $b_0/\sqrt{2} < a < b_0$, i.e., the arc observed is approximately the size of the cosmic string. Using the "empty beam" model with $\Omega_0 = 1$, the following relation is obtained:

$$\frac{b_0 H_0}{8\pi d(Z_d, Z_s)} < \mu < \frac{\sqrt{2} b_0 h_0}{8\pi d(Z_d, Z_s)}$$

or

$$4 \times 10^{-6} < \mu < 6 \times 10^{-6} \quad ,$$

where $Z_d = 0.373$, and $Z_s = 0.724$.

Discussion. We know from the present isotropy measurement of the microwave background that $\mu < 10^{-5}$, and μ cannot be lower than 10^{-6} if galaxy formation was assisted by the cosmic string. This is very consistent with the result obtained in this paper. It is interesting to see that the observed loop radius, a, is roughly of the size of the smallest loop, μH_0^{-1}. This means that the cosmic string might exist in the form of the smallest possible loop in the present universe (Mitchell and Turok 1987).

It is important to distinguish two kinds of deflectors: cosmic strings and clusters of galaxies. The arc image due to the cosmic string should lie outside the loop string, and the radius of arc should not be less than the radius of the string, which is about 50 kpc. The arc image caused by the cluster of galaxy could be of any size, in principle. The radii of three samples of luminous arcs obtained so far are larger than or approximately equal to 50 kpc. Larger samples are needed to determine which type of deflector causes the arcs.

The preliminary theory of the statistical properties of cosmic strings shows that for every arc with length $L \geq 20''$, there would be about four medium-sized arcs with lengths $L \approx 10'' \sim 20''$. However, the observations do not prove this. The same problem also appeared in the model in which a cluster of galaxies is the deflector. The medium-sized arcs may not have been detected because of selection effects.

Acknowledgments. I wish to thank Prof. Lizhi Fang, Prof. Zhenglong Zou, and Prof. Jiansheng Chen for useful discussions during the preparation of this paper.

References.

Gott, J. R. 1985, *Ap. J.*, **288**, 422.

Grossman, S. A., and Narayan, R. 1988, *Ap. J. (Letters)*, **324**, L37.

Lavery, R. J., and Henry, J. P. 1988, *Ap. J. (Letters)*, **329**, L21.

Lynds, R., and Petrosian, V. 1986, *Bull. AAS*, **18**, 1014.

Miller, J. S., and Goodrich, R. A. 1988, *Nature*, **331**, 685.

Mitchell, D., and Turok, N. 1987, *Phys. Rev. Letters*, **58**, 1577.

Paczynski, B. 1987, *Nature*, **325**, 572.

Soucail, G., *et al.* 1987*a*, *Astr. Ap.*, **172**, L14.

Soucail, G., *et al.* 1987*b*, *IAU Circ.*, No. 4482.

VI. Surveys, Searches, and Statistics

RESULTS OF THE VLA GRAVITATIONAL LENS SURVEY

J. N. Hewitt[1], B. F. Burke[2], E. L. Turner[3], D. P. Schneider[4], C. R. Lawrence[5], G. I. Langston[6], and J. P. Brody[2]

[1]Haystack Observatory, Westford, MA 01886
[2]Department of Physics, Massachusetts Institute of Technology, Cambridge, MA 02139
[3]Princeton University Observatory, Peyton Hall, Princeton, NJ 08544
[4]School of Natural Sciences, Institute for Advanced Study, Princeton, NJ 08540
[5]Mail Stop 105-24, California Institute of Technology, Pasadena, CA 91125
[6]National Research Council Associate, Naval Research Laboratory, Code 4131, Washington, DC 20375-5000

Introduction. On this occasion of Bernard Burke's birthday, it seems appropriate to assess the status of a project for which he has supervised radio observations for a number of years. This project is a gravitational lens search based on a large survey (Lawrence *et al.* 1984a, Bennett *et al.* 1986) carried out with the National Radio Astronomy Observatory* Very Large Array (VLA). Observations carried out at the VLA with the express purpose of searching for gravitational lenses began in 1983, and approximately 4000 sources were observed with the VLA from 1981 through 1986. These observations allow results of a statistical nature to be derived, as well as identification of new lens systems. A number of promising lens candidates have been discovered, of which the lens interpretation of three seems quite compelling. Rigorous verification or elimination of the lens hypothesis is time consuming, involving optical observations and multifrequency radio observations. We review here some statistical results of the VLA survey and the published candidate lens systems that have been discovered. We also present preliminary results of a new candidate lens system.

The Survey. All sources observed in the VLA lens search were drawn from the MIT-Green Bank (MG) survey (Bennett *et al.* 1986) and from preliminary analysis of the MG survey data (Lawrence *et al.* 1986). The MG survey, carried out with the NRAO 300-foot telescope in Green Bank, West Virginia, covers essentially the entire sky from $-0.5°$ to $+19.5°$ in declination, excluding the galactic plane. The observing frequency was 4.8 GHz, giving a beamwidth of 3 arc minutes; the completeness flux density limit ranges from 53 mJy to 106 mJy. The positional accuracy derived from the MG survey (approximately 30 arc seconds) is insufficient for reliable

* The National Radio Astronomy Observatory is operated by Associated Universities, Inc., under contract with the National Science Foundation

optical identification of the radio sources, but is within the field of view of VLA A-array observations at 5 GHz. The MG survey source list consists of 5974 radio sources detected with a signal-to-noise ratio of 5 or greater, and a supplemental list of 3836 sources detected with a signal-to-noise ratio between 4 and 5.

Observations with the VLA of the MG survey sources took place during 11 observing runs in May 1981, February 1982, September and October 1982, September 1983, December 1984 (two observing runs), February 1985 (three observing runs), and April 1986 (two observing runs). The motivation for the first three runs was a statistical study of extragalactic radio sources; sources over a range of flux density bins (from 50 mJy to 500 mJy) were selected for observation, and some were selected by spectral index. Unfortunately, for the purposes of the gravitational lens search that benefits from high resolution, two of these early observing runs were carried out when the VLA was in the B-array (resolution $\sim 1.2''$). The last eight observing runs were carried out with the VLA in the A-array (resolution $\sim 0.4''$). For the lens search observations, sources were selected by flux density and galactic latitude only; the brightest MG sources outside the galactic plane were observed. Except for a few sources that were missed because of problems with the VLA, the VLA observations are complete to 115 mJy.

The selection of gravitational lens candidates from the sources mapped with the VLA is based on two considerations. First, it is unusual for radio sources to display more than one compact component due to structure intrinsic to the source (though not, of course, impossible; for example, unresolved hot spots and double nuclei of galaxies have been observed). Gravitational lensing can produce double images of the core of a radio source, and the core is likely to have an optical counterpart for which the optical spectrum can be measured, providing a means to verify the lensing interpretation. Second, gravitational lensing produces distortions in extended sources (though, again, distortions may be due to other processes, such as interaction of a jet with a dense intergalactic medium). The radio sources were ranked by dividing them into three groups: (1) those with at least two unresolved components, (2) those with at least two components, not necessarily unresolved, and (3) single unresolved sources. Within the first two groups, sources showing distortion (i.e., noncollinearity) were given higher priority. From this ranking, sources are being selected for optical followup observations. Multiple radio components with optical counterparts at the same redshift with similar spectra are likely to be gravitationally lensed.

<u>Statistical Results.</u> Samples of sources can be used to measure the statistics of gravitational

lensing and provide information on cosmological distributions of matter. Theoretical relationships between gravitational lens observables and the properties of populations of lenses of various types have been worked out (cf. Press and Gunn 1973; Canizares 1982; Turner, Ostriker, and Gott 1984; Dyer 1984), and there have been some attempts to relate these calculations to available data sets (Press and Gunn 1973; Canizares 1982). A subsample of the data from the VLA lens survey was reanalyzed to bring all the VLA maps to a uniform level of quality (dynamic range of 17:1 or better), and the expected observable properties of lens images were calculated taking explicitly into account the instrumental limits of the VLA. From a count of all the unresolved (i.e., not lensed on scales detectable with the VLA) sources in the subsample and a comparison with the calculations, a very conservative upper limit was placed on the cosmological distribution of lenses. In particular, we conclude that the universe cannot be closed with a population of compact masses in the range of 10^{11} to 10^{12} solar masses (Hewitt 1986). The properties of cosmological populations of lenses can be much more severely constrained by the VLA data by (1) taking into account observable quantities other than just the frequency of lensing, such as image separations and flux density ratios, (2) using the measured redshift distribution of the sources, rather than conservatively assuming the redshifts are relatively small, and (3) including the effects of flux density intensification bias (cf. Turner, Ostriker, and Gott 1984) in the calculations. Work is in progress in these areas.

Candidate Lens Systems. The above limit on the number of massive compact objects is based on an assessment of the number of sources in the VLA sample that *are not* lensed. The VLA survey has also discovered sources that appear to be lensed. We review the properties of the three published systems here: 2016+112, 0023+171, and MG1131+0456. Of these three, the evidence for lensing in 2016+112 and MG1131+0456 is quite strong; the evidence for 0023+171 is less convincing, and the system could represent a binary galaxy pair. We also present data on a new candidate lens system, MG0414+0534.

2016+112 was discovered during the first VLA observing run (Lawrence *et al.* 1984b). At radio wavelengths it appeared as three unresolved (at the resolution of the original observations, but see Langston *et al.*, this volume) radio sources arranged in a right triangle (Fig. 1). The strong indication that 2016+112 is lensed came from optical spectra of components A and B. The spectra show strong, unusually narrow, emission lines at the same redshift (velocity difference < 40 km/sec). Since its discovery, 2016+112 has been studied further in the radio and the optical, and it is recognized to be an extremely complex system. Broad band optical imaging has revealed two foreground galaxies, one near the position of radio source C and another

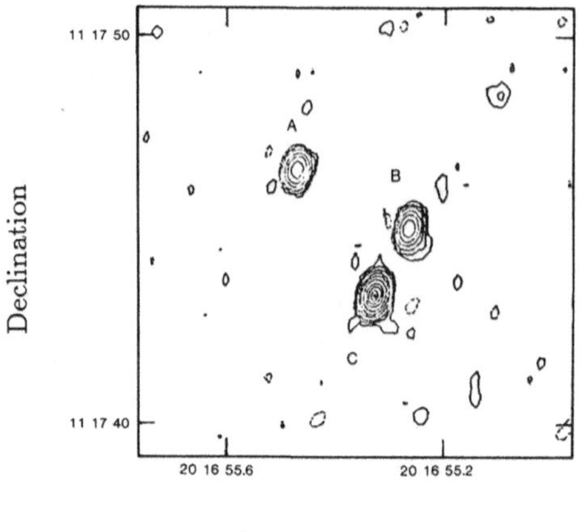

Fig. 1. Contour plot of VLA image (λ 6 cm, A array) of 2016+112. Reproduced from Lawrence *et al.* (1984b).

(named D) near the center of the A-B-C triangle (Schneider *et al.* 1985). There is no evidence for a cluster in the field. Through narrow band optical imaging, Lyman α emission at the quasars' redshift was detected near radio source C, indicating that C is a composite of a radio galaxy and the third image of the quasar. In addition, two diffuse narrow-line emission regions, which appear to be physically distinct, are located near quasar images A and B (Schneider *et al.* 1986, 1987). The interpretation of 2016+112 as a lensed system seems secure, given the spectral similarity of the quasar images and the presence of three images and of galaxies C and D. Models have been found that reproduce fairly well the observed properties of 2016+112. These models consist of lensing matter at the positions of the two galaxies and in the form of a large pool (velocity dispersion of approximately 1000 km/sec) of matter associated with a surrounding (unseen) cluster. The models require the presence of the smoothly distributed mass associated with the cluster, indicating there may be dark matter in the 2016+112 system (Narasimha, Subramanian, and Chitre 1987). To determine whether dark matter *must* be invoked to explain the imaging, a systematic search for lensing models that require matter associated with the galaxies only should be carried out.

The second lens candidate discovered in the VLA survey is 0023+171 (Hewitt *et al.* 1987). The radio structure of 0023+171 is shown in Figure 2, and the two optical counterparts in the field are indicated. Components A and B and the southern optical counterpart display the common structure of two radio lobes emanating from an unresolved optical core; component C and the northern optical counterpart are both unresolved. The evidence for lensing is in the similarity of the properties of the optical counterparts. Both show emission lines, [O II]

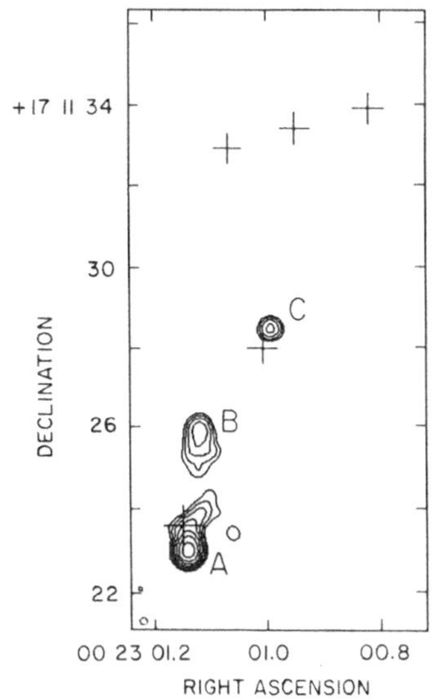

Fig. 2. Contour plot of VLA image (λ 6 cm, A array) of 0023+171. Positions of optical objects in the field are indicated by crosses. Reproduced from Hewitt *et al.* (1987).

and [Ne III], at the same redshift and the flux density ratios in the lines are not significantly different. In contrast, a study of the optical spectra of radio sources at similar redshifts showed that the flux density ratios varied over two orders of magnitude in 11 sources in which the same lines were detected (Allington-Smith, Lilly, and Longair 1985). The difference in the radio and optical flux density ratios and the pointlike nature of C at radio wavelengths, however, are not explained by a simple (i.e., a single lens) model. Thus, the lensing interpretation of 0023+171 is not very secure. If it is lensed, however, it is potentially an interesting system as lensing would require a substantial amount of dark matter.

The third lens candidate of the VLA lens search is the remarkable system MG1131+0456. We describe here in chronological order the observations that have led to the conclusion that MG1131+0456 is probably gravitationally lensed. The first observation of MG1131+0456 was a VLA snapshot on 17 December 1984; even with the relatively poor quality of these data, the smooth ringlike structure was evident (see Fig. 3). An "Einstein ring" gravitational lens seemed a possible explanation, but other astrophysical systems, such as H II regions, planetary nebulae, and supernova remnants, display ringlike morphology. The nonthermal radio spectral index of 0.9 ± 0.2 (calculated from published flux densities at 1.4 and 5 GHz (Niell 1971; Bennett *et al.* 1986) ruled out the H II region and planetary nebula interpretations. An optical (R band) image of the field was acquired in March 1987 with the Kitt Peak 4-meter telescope. In seeing of $1\overset{''}{.}1$, a faint ($m_R \cong 22$) object was detected; the object is elliptical with a position angle approximately

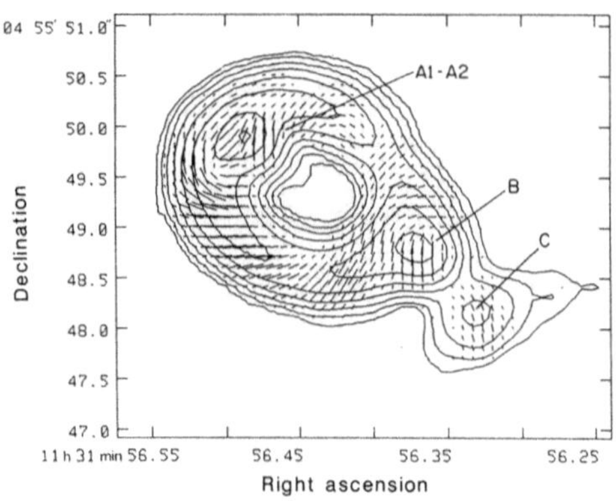

Fig. 3. Contour plot of VLA image (λ 6 cm, A array) of MG1131+0456. The line segments represent the intensity and position angle of linearly polarized radiation. Reproduced from Hewitt *et al.* (1988). Reprinted from *Nature*, vol. 333, p. 537. Copyright © MacMillan Journals Limited.

equal to that of the ring. A spectrum of the optical counterpart was measured in April 1987, also with the Kitt Peak 4-meter telescope. A continuum with no emission lines was detected, indicating that MG1131+0456 is not likely to be a supernova remnant. The radiative properties are typical of a radio galaxy, and the unusual structure is precisely what one would expect of a gravitational lens. A plausible imaging geometry is a lens with elliptical symmetry lying in front of a radio source that consists of a core and two lobes. The lobe directly behind the lens is imaged into a ring; the core is close enough to the lens to form two bright images; and the second lobe is far enough away that only one bright image of it is formed. The second lobe is evident in Figure 3, and the geometry of the ring and the cores is clearly seen in the λ 2 cm map of Figure 4. Supporting this interpretation is the fact that the angle of linear polarization of the

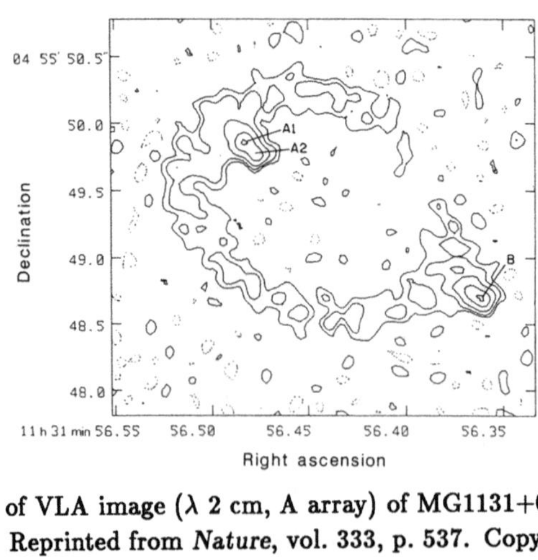

Fig. 4. Contour plot of VLA image (λ 2 cm, A array) of MG1131+0456. Reproduced from Hewitt *et al.* (1988). Reprinted from *Nature*, vol. 333, p. 537. Copyright © MacMillan Journals Limited.

two cores is the same (though dual-frequency polarization measurements should be done to measure any effects of Faraday rotation). Simulations of lensing by a potential with elliptical symmetry (Kochanek *et al.* 1988; Blandford, this volume) reproduce the geometry and polarization pattern of MG1131+0456 in considerable detail.

The strongest evidence supporting the lensing interpretation of MG1131+0456 is the extremely unusual geometry of the system, the inconsistency of its radiative properties with those of other known ringlike objects, and the consistency with gravitational lensing of the pattern of linear polarization. There are, as yet, no data in conflict with the lensing interpretation; however, one can never rule out the possiblity of a new type of astronomical object.

The probability of observing a nearly closed Einstein ring in the VLA survey, which unlike optical quasar surveys includes many *extended* sources, is not unreasonable. A numerical calculation for an elliptical (ellipticity = 0.3) gravitational potential has been carried out to compare the probability of producing a ring image of an extended source to the probability of producing two images of a point source. For lens and source redshifts of 0.5 and 1.0, the source radius must be 0.″8 for the two probabilities to be equal (Hewitt *et al.* 1988). This is comparable to the size of the lobes of extragalactic radio sources and, since approximately half the sources in the VLA survey are resolved, it would not be surprising to find several Einstein rings among the survey sources. In fact, there is strong indication that another source has ringlike morphology due to lensing (Langston *et al.*, in preparation).

More tests of the lensing of MG1131+0456 are possible and should be carried out. Some possiblities are a comparison of the VLBI structure of the core components, a search for two redshift systems in the optical spectra, and a comparison of the light curves of the core components, including evidence for a time delay. Work on these tests is in progress or planned, and we report a preliminary VLBI result here. MG1131+0456 was observed for 719 seconds at λ 18 cm in November 1987 with a VLBI array consisting of the Haystack 36-meter, the Green Bank 43-meter, the Fort Davis 26-meter, and the phased VLA telescopes. Correlation was carried out on the Haystack Observatory's Mark III correlator, and the source was detected on the most sensitive (VLA-Green Bank) baseline with fringe spacings of 2 mas (north-south) and 7 mas (east-west). The detection was at the fringe delay and rate expected for the B compact component and the correlated flux density was 3 mJy. The expected peak VLA flux density of component B, assuming a spectral index of 0.4, is 6 mJy. These results show that there is compact structure in at least one of the core components, and VLBI tests of the lensing interpretation of MG1131+0456 are possible. Mapping of the core components on milli-arc second scales would allow a comparison of the relative magnification of the images to models, and, if there is motion in a compact component, a measurement of the time delay through monitoring the structure.

In summary, the evidence supporting the lensing interpretation of MG1131+0456 is quite strong, though more tests can, and should, be carried out. MG1131+0456 is potentially important to lens studies because of the presence of extended structure completely surrounding the gravitational potential that should serve to tightly constrain models of the mass distribution in the lens.

We now present preliminary results of a new lens candidate, MG0414+0534. The source was first observed on 17 December 1984 as part of the VLA survey, and the unusual radio structure in the A-array λ 6 cm map (Fig. 5) made it a high priority lens candidate. B-array λ 2 cm data, with resolution similar to the survey data, were acquired in December 1987. Figure 6 shows the λ 6 cm and the λ 2 cm maps convolved to the same resolution; the agreement between the two is remarkable. Fitting four Gaussian components to the radio map of Figure 5(b) results in flux densities of 360, 320, 130, and 50 mJy. Components A, B, and C are resolved. Figure 7 shows a contour plot of an R-band image of MG0414+0534 acquired with the Kitt Peak 4-meter telescope in September 1987 in seeing of $1''.0$. There clearly are three components in the optical image, and the brightest is slightly resolved, indicating that it may be

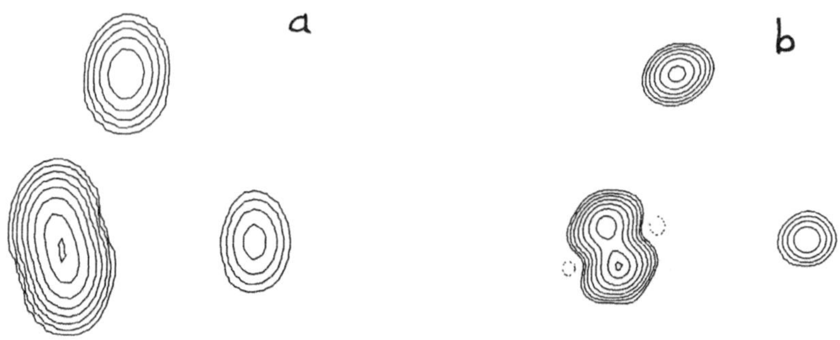

Fig. 5. (a) Contour plot of VLA image (λ 6 cm, A array) of MG0414+0534 (resolution $= 0''.5$). (b) Contour plot of VLA image constructed from the same data as (a), but with clean components convolved with a $0''.25 \times 0''.25$ clean beam.

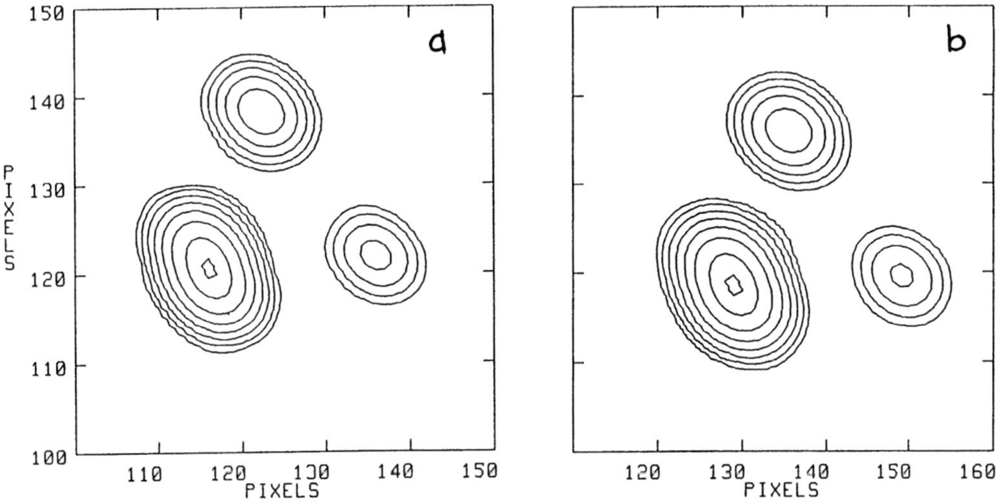

Fig. 6. (a) Contour plot of VLA image (λ 2 cm, B array) of MG0414+0534. (b) Contour plot of image of Fig. 5(a), convolved to the same resolution as Fig. 6(b).

Fig. 7. Contour plot of optical (R band) image of MG0414+0534.

double. The light in the image is entirely described by four point-spread functions with the geometry of the radio components; there is no evidence for light from a lensing galaxy. The 5 GHz flux densities of the three components (A and B combined, C, and D) are in the ratio of 15:3:1; the 15 GHz flux densities are in the ratio of 14:3:1; and the optical fluxes are in the ratio of approximately 10:2:1. The optical fluxes are poorly determined because of the seeing, but are probably consistent with the radio measurements. A spectrum of the combined light from components A, B, and C has been measured with the Palomar 5-meter telescope. The light is unusually red, and there is no evidence for emission lines.

The four-image configuration and relative brightnesses of the components give the imaging expected from a gravitational potential with elliptical symmetry (Grossman and Narayan, this volume). Detailed modelling of MG0414+0534 using the elliptical potential parameterized by Blandford and Kochanek (1987) reproduces the positions of the four components to within $0''.2$ (Hewitt *et al.*, in preparation).

The case for MG0414+0534's being a lens is quite compelling because of the radio and optical morphology and the agreement between the flux density ratios at both radio and optical wavelengths. The optical spectroscopy is inconclusive because of the poor spatial resolution and lack of sharp features in the spectrum. MG0414+0534 is by far (by nearly a factor of 10) the brightest lens candidate at radio wavelengths, and VLBI mapping of the components to check the similarity of their structure is strongly indicated. Preliminary VLBI observations are scheduled in September 1988. These should determine whether the components of MG0414+0534 have compact structure detectable on VLBI scales and may give some indication of the components' structure.

Summary. Approximately 4000 sources have been observed with the VLA, and these sources are being examined for evidence of gravitational lensing, both to determine the statistical properties of gravitational lensing and to find new lens systems. From just an upper limit on the frequency of lensing in a subsample of the VLA data, one can show that the universe cannot be closed by a uniform population of 10^{11} to 10^{12} solar mass compact objects. The VLA survey has also resulted in a number of promising lens candidates, sources with unusual structure with multiple radio-optical counterparts. We have discussed the four best studied cases: 2016+112, 0023+171, MG1131+0456, and MG0414+0534. The spectral similarity of the components of 2016+112 and the presence of three images and two foreground galaxies makes the lensing interpretation of 2016+112 quite secure. 0023+171 shows the classic lens signature of nearly identical optical spectra of the components, but the complex radio morphology is difficult to understand within the context of a simple lensing model. Further data and modelling are necessary to determine whether 0023+171 is in fact lensed. If it is, however, the large mass-to-light ratio implied for the lens makes it a particularly interesting system. In MG1131+0456 and MG0414+0534, no lensing galaxy has been detected, nor have separate optical spectra of the components been measured. Rather, the evidence for lensing comes from the unusual morphology that reproduces well that expected for lensing, and other radio data: the polarization angles in the case of MG1131+0456, and the spectral indices in the case of MG0414+0534. In both cases, there is more structure than simply two stellar objects, potentially providing better constraints on lens models than have been achieved in the past. Also in both cases, there appears to be insufficient light in the field for the lensing to be due to a "normal" galaxy, an intriguing fact that at the same time raises the possiblity of studying dark matter and demands more rigorous proof of lensing.

References.

Allington-Smith, J. R., Lilly, S. J., and Longair, M. S. 1985, *M. N. R. A. S.*, **213**, 243.

Bennett, C. L., Lawrence, C. R., Burke, B. F., Hewitt, J. N., and Mahoney, J. H. 1986, *Ap. J. Suppl.*, **61**, 1.

Blandford, R. D., and Kochanek, C. S. 1987, *Ap. J.*, **321**, 658.

Canizares, C. R. 1982, *Ap. J.*, **263**, 508.

Dyer, C. C. 1984, *Ap. J.*, **287**, 26.

Hewitt, J. N. 1986, Massachusetts Institute of Technology Ph.D. Thesis.

Hewitt, J. N., Turner, E. L., Lawrence, C. R., Schneider, D. P., Gunn, J. E., Bennett, C. L., Burke, B. F., Mahoney, J. H., Langston, G. I., Schmidt, M., Oke, J. B., and Hoessel, J. G. 1987, *Ap. J.*, **321**, 706.

Hewitt, J. N., Turner, E. L., Schneider, D. P., Burke, B. F., Langston, G. I., and Lawrence, C. R. 1988, *Nature*, **333**, 537.

Kochanek, C. S., Blandford, R. D., Lawrence, C. R., and Narayan, R. 1988, *M. N. R. A. S.*, in press.

Lawrence, C. R., Bennett, C. L., Hewitt, J. N., and Burke, B. F. 1984a, *Ap. J. (Letters)*, **278**, L95.

Lawrence, C. R., Bennett, C. L., Hewitt, J. N., Langston, G. I., Klotz, S. E., Burke, B. F., and Turner, K. C. 1986, *Ap. J. Suppl.*, **61**, 105.

Lawrence, C. R., Schneider, D. P., Schmidt, M., Bennett, C. L., Hewitt, J. N., Burke, B. F.,

Turner, E. L., and Gunn, J. E. 1984b, *Science*, **223**, 46.

Narasimha, D., Subramanian, K., and Chitre, S. M. 1987, *Ap. J.*, **315**, 434.

Niell, A. 1971, Cornell University Ph.D. Thesis.

Press, W. H., and Gunn, J. E. 1973, *Ap. J.*, **185**, 397.

Schneider, D. P., Gunn, J. E., Turner, E. L., Lawrence, C. R., Hewitt, J. N., Schmidt, M., and Burke, B. F. 1986, *A. J.*, **91**, 991.

Schneider, D. P., Gunn, J. E., Turner, E. L., Lawrence, C. R., Schmidt, M., and Burke, B. F. 1987, *A. J.*, **94**, 12.

Schneider, D. P., Lawrence, C. R., Schmidt, M., Gunn, J. E., Turner, E. L., Burke, B. F., and Dhawan, V. 1985, *Ap. J.*, **294**, 66.

Turner, E. L., Ostriker, J. P., and Gott, J. R., III. 1984, *Ap. J.*, **284**, 1.

OPTICAL SEARCHES FOR GRAVITATIONAL LENSES

Rachel L. Webster
Canadian Institute for Theoretical Astrophysics
60 St. George Street, Toronto, Ont., Canada, M5S 1A1

Paul C. Hewett
Institute of Astronomy
Madingley Road, Cambridge, UK, CB3 0HA

Abstract. Although individual occurrences of gravitational lensing can provide some constraints on the mass distribution in the deflecting object, statistical samples of lensed quasars are necessary to provide a fuller description of the overall mass distribution in the universe. To date, optical searches for multiply imaged quasars have been hampered by poor resolution ($\gtrsim$ 2–3″), unknown selection biases in the determination of the quasar samples, and relatively small numbers of objects. However, dramatic improvements in all these areas are now possible and a new series of optical gravitational lens searches are underway offering the potential for a large increase in the number of bona fide gravitational lensing systems over the next few years. This paper summarizes the status of optical gravitational lensing surveys and reviews the problems specific to this type of search. The techniques and preliminary results from the Gravitational Lens Survey from the Automatic Plate Measuring (APM) Facility are used to illustrate the potential of the new generation of surveys. Some applications of the lens samples now being compiled are discussed.

1. Introduction. Zwicky (1937) made a reasonable estimate of the expected frequency of multiply imaged extragalactic objects; however, the discovery of the first gravitational lens occurred serendipitously some 42 years later (Walsh *et al.* 1979). Further examples were discovered by chance, but early systematic searches produced disappointingly low success rates. The task of locating gravitational lenses is not easy: if galaxies are the predominant lensing population, the spatial separations of multiply imaged optical quasars will be $\sim 1''$ and only detectable by imaging studies conducted in very good seeing (Turner 1980). These considerations prompted Hewitt, Turner, and collaborators (Hewitt *et al.* 1988b) to instigate a radio survey sensitive to subarcsecond separations, for objects that might be lensed quasars or other extragalactic objects. Several strong candidates for gravitational lenses have been discovered, though as yet none has the separations of $\lesssim 1''$ anticipated at the onset.

If the mass attributable to galaxies and galaxy clusters from dynamical arguments represents most of the clumped mass component in the universe, the instance of multiple imaging in the overall quasar population might be as low as $\sim 10^{-3}$ (Turner *et al.* 1984). Fortunately, a bias operates to preferentially include amplified lensed quasars in any flux-density-limited sample. The size of this bias is particularly sensitive to the slope of the intrinsic quasar luminosity function, and of course to Ω_{lens}, the mean mass density in objects capable of significant amplification. Thus, bright magnitude-limited samples of quasars will contain a higher fraction of lensed images.

Automated surveys for quasars and gravitational lenses over large areas of sky to faint magnitudes are now possible, and optical imaging at good sites and large platescales makes the detection of images with separations $\sim 0.5''$ possible. Thus, surveys that take advantage of the amplification bias and that are sensitive to the separations expected from "galaxy-sized" clumps are viable in the optical. A statistical sample of gravitationally lensed quasars can be used to determine the mass function of the clumped component of matter in the universe. One of the great strengths of such an approach is that it provides an independent method for determining both the space density and some characteristics of the potentials responsible for lensing. Other techniques for the determination of the large-scale mass distribution rely on assumptions about the mass-to-light ratios and/or the dynamical state of the systems, neither of which is well constrained. A range of other problems may be explored using gravitational lensing, including: modelling mass distributions in individual lenses, measurement of the Hubble constant, determination of the geometric structure of the central parts of quasars and quantifying perturbations in all cosmological measurements.

Section 2 discusses some of the problems inherent in optical surveys that are capable of providing statistical samples of gravitational lenses. Section 3 lists the optical surveys currently underway. The first results of the APM Gravitational Lens Survey are discussed in section 4. Finally, some applications of these new statistical samples are outlined.

2. What Should an Optical Search Do? A key role of the large optical surveys will be to provide a statistically complete sample of gravitationally lensed quasars. The sample must be large enough for comparison with theoretical predictions, while also being compatible with the restrictions imposed by limited telescope time; thus, a sample of 10–20 lenses is a realistic aim. Surveys at all wavelengths are subject to limitations in their ability to detect multiply imaged quasars as a function of component angular separation, component magnitude difference and

overall system brightness. "Completeness" in this context does not, therefore, mean the ability to detect all instances of gravitational lensing. It is more important that the survey be sensitive to a substantial dynamic range in each of the properties describing the appearance of a multiply imaged quasar, and most important that the dependence of the probability of detection as a function of each property be precisely specified. A simple, but often ignored example is the derivation of the flux limit and associated dispersion determining the inclusion (or otherwise) of quasars in the sample to be surveyed for evidence of gravitational lensing. A more complex factor is the determination of the survey's sensitivity to lenses as a function of component magnitude differences and angular separation. Provided that digital data is available, careful computer simulation combined with a precise description of the observational configuration allows detailed "detection functions" to be derived (e.g., Webster *et al.* 1988a).

There are four principal characteristics of images that are signatures of gravitational lensing: (i) multiple images of the same object, (ii) a background image seen either as a nearly complete ring (Hewitt *et al.* 1988a) or as an extended arc (Lynds and Petrosian 1986, Soucail *et al.* 1987, Lavery and Henry 1988), (iii) a statistical excess of quasars or other background sources associated with some population of lensing objects at intervening redshifts – examples might include quasars that are juxtaposed with nearby galaxies, or that show an excess of certain types of absorption systems in their spectra, (iv) images exhibiting brightness variations characteristic of compact objects crossing the line of sight (e.g., Kayser *et al.* 1986).

The first two image configurations are evidence of mass concentrations where the mean surface density between the multiple images or within the arc is in excess of the critical value. The critical surface density is defined as $\Sigma_{crit} = D_{OS}/4\pi D_{OL} D_{LS}$, where the subscripts of the distances refer to the observer, lens, and source. Observed occurrences of these mass concentrations have angular scales of $\gtrsim 1''$. In instances where multiple images cannot be resolved, such as for configurations (iii) and (iv), it is not possible to distinguish between quasars that are microlensed and those that are amplified by a subcritical surface density. Microlensed images are sensitive to compact objects with masses

$$M \gtrsim 10^{-5} \left(\frac{r_s}{10^{15}\ \mathrm{cm}}\right)^2 \left(\frac{10^3\ \mathrm{Mpc}}{\mathcal{D}}\right) M_\odot$$

where r_s is the source radius, and $\mathcal{D} = D_{OD}/D_{OS} D_{DS}$. Microlensing actually results in two images whose separations are $\sim 10^{-6''}$, but such separations will remain impossible to detect directly in the optical for the foreseeable future. Unlike microlensing, where the compact objects are effective lenses at any redshift, the effectiveness of an object with a subcritical surface density varies strongly with lens redshift. Thus, to produce a significant effect, the lensing population must have surface densities in a rather well-defined range, and the effects will be strongly dependent on the observer-lens/lens-source distance ratio.

It is important to emphasize that the direct effects of gravitational lensing will only identify the "tip of the iceberg" – the very clumped component. More smoothly distributed mass will also produce effects through the shape of the large-scale metric. In addition, many weak lenses can combine to significantly affect a background source (Dyer and Oattes 1988).

Searches for complete samples of gravitational lenses face a number of problems, some of which are specific to optical work, and others that bedevil all wavelengths.

(i) The dependence on image morphology of the selection algorithms used to define the base quasar catalogue must be well determined. It is extremely difficult to make use of the large, well known compilations of quasars for statistical work involving multiple images with separations exceeding $\sim 1.5''$, or for quasar-galaxy juxtapositions with separations comparable to, or smaller than the galaxy diameters.

(ii) Critical requirements for statistical analysis are a well-defined flux limit and dispersion for the sample, together with a well-defined search area. A specific knowledge of the quasar luminosity function several magnitudes fainter than the magnitude limit of the survey is also required in order to determine the effects of amplification bias.

(iii) Compared with radio surveys, the greatest limitation of the optical searches has been resolution. Since image separations of $\sim 1''$ are predicted for galaxy-sized lenses, only optical observations made in the very best seeing conditions provide the necessary resolution. However, it has been ably demonstrated by Surdej *et al.* (1988) that the required observations can be made by ground-based telescopes.

(iv) Obtaining a large enough sample to undertake statistical calculations is a prime consideration. Estimating numbers is difficult since Ω_{lens}, the main factor required, is precisely what we don't know. If we undertake reasonable estimates for quasars with $m_B \lesssim 18.5$, then the total number of multiply imaged quasars in the sky would be

$$\sim 600 \left(\frac{P_{miq}}{0.01}\right)\left(\frac{\rho_{qso}}{1.5\ \text{sq. deg.}^{-1}}\right)$$

or

$$\sim 1.5 \times 10^{-2} \left(\frac{P_{miq}}{0.01}\right) \left(\frac{\rho_{qso}}{1.5}\right) \quad \text{per square degree}$$

where P_{miq} is the probability of a quasar being multiply imaged (which is a strong function of the magnitude limit and Ω_{lens}), and ρ_{qso} is the surface density of quasars to $m_B = 18.5$. The analysis of 30–50 Schmidt fields to a magnitude limit of $m_B \lesssim 18.5$ should yield a sufficient sample for theoretical calculations. A survey based on the present X-ray selected samples is not viable due to the low surface density, small absolute numbers and bias towards redshifts $\lesssim 1$ (e.g., Maccacaro *et al.* 1984). The number density of radio quasars is also much lower than optical quasars, with only $\sim 10\%$ of optical quasars detectable in a "survey mode" at radio wavelengths. Hewitt *et al.* (1988a) have demonstrated that such a search will find useful gravitational lens candidates. Compilation of a statistical sample of lensed radio sources may, however, be difficult since the optical counterparts of candidate lenses are often very faint, making spectroscopy difficult. Also the luminosity function of faint radio quasars is poorly known.

(v) The final issue is "verification" or "when is a lens a lens?" This issue has become more pressing with the discovery of several binary quasars (Djorgovski *et al.* 1987, Crampton *et al.* 1988). Estimates of the frequency of close pairs give the probability of random juxtaposition for two quasars as small (Phinney and Blandford 1986), unless the quasars are highly correlated. If the quasar two-point correlation function has the same slope as the galaxy two-point correlation function, but is perhaps a factor of two stronger (Shanks *et al.* 1987), then using the formulation of Phinney and Blandford, we calculate that the probability of two correlated quasars with $\theta < 2''$, $m < 21$ and $z \sim 2$ is $\sim 5 \times 10^{-4}$. This result suggests that one such pair might be found in every 1000 quasars. This probability takes no account of spectral information, and assumes that the correlation function can be extrapolated to proper distances of $\sim 20\ h^{-1}$ kpc.

Criteria used to identify systems as gravitational lenses have included identical redshifts, similar emission and absorption spectra, identification of the lensing object, measurement of a time delay, more than two images, identical ratios between images at optical and radio wavelengths, similar polarization, lensing of other background objects and similarities in the shapes of the emission lines. Our lack of knowledge concerning the sizes of, and geometrical relation between, the regions responsible for generating the optical continuum, the emission lines and radio continuum, means that simple arguments that favor a very close similarity between the properties of the different components in a lens may not be correct. In practice, once certain basic criteria, such as very similar redshifts, have been met, the assessment of whether two or more quasar images are of one object rests on probability arguments based on our knowledge of the incidence of close physical quasar pairs.

A successful optical survey requires both a means of identifying quasars coupled with information on direct image properties. Quasars may be identified from multicolor (including UVX searches) or slitless spectroscopic material. Other techniques are unlikely to achieve a sufficiently high surface density to be useful. Direct image information to a considerably fainter limiting magnitude should also be available, otherwise a severe restriction is placed on the magnitudes of secondary images that may be detected. The requirement that a detailed knowledge of the detection function is available excludes visual searches and makes digital data, either directly from CCD scans, or through machine scanning of photographic plates, a prerequisite. Once the analysis of the survey data has been completed, higher resolution spectroscopy and deep CCD imaging of a candidate list is required.

3. Surveys for Gravitational Lenses. Table 1 lists various groups that are in the process of completing, or have completed optical surveys for gravitational lenses. In addition, a number of workers are compiling complete samples of quasars, which could be used for lens studies (see Foltz and Osmer 1988, for a summary).

In the most successful optical survey to date, Surdej and collaborators (Surdej *et al.* 1988) have imaged a sample of 111 quasars selected from the Veron catalogue (Véron-Cetty and Véron 1987) to have $m_v \lesssim 18.5$ and $M \lesssim -29$ at high resolution. The seeing for these observations was $\sim 1''$. Two objects have been verified as multiply imaged quasars, three more are good candidates, and a further 20 show associated structure in the form of fuzz, a faint nearby galaxy, a jet or some other feature. The separations of the two confirmed candidates are $2.2''$ and $1.36''$. This survey has already shown that many high-luminosity quasar images have structure, though whether this is due to lensing or to detection of objects at the same redshift as the quasars remains to be established. From these data, $P_{miq}(\gtrsim 1'') \sim 2$–5%.

Weedman and Djorgovski (1988) have visually searched 200-400 quasars to a limit of $m_B \sim 20.5$–21 on the best available CFHT grens plates. The seeing is estimated as 0.8–1.2$''$ (FWHM). They find no candidates for multiply imaged quasars with separations greater than $2''$. This result provides a 3σ upper limit of $3/N_q$ or $P_{miq}(\gtrsim 2'') = 0.8$–1.5%. This result should be compared with the statistics of lenses found in the samples of Sramek and Weedman (1978) and Gaston (1983), where two multiply imaged quasars were identified in a sample of

Table 1: Optical Surveys for Lensed Quasars

Group	Status	Data	Scanning	m_{lim}	Numbers
Borra *et al.*	preliminary	CFHT grens	PDS	$m_B < 20 - 21$	–
Crampton *et al.*	preliminary	CFHT grens	eye	$m_B < 20.5$	~ 1000
Djorgovski and Meylan[1]	in progress	Hewitt & Burbidge Catalogue[2]	various	$M_v < -28$ $z > 1.5$	$> 200 - 250$
Engels *et al.*	preliminary	Schmidt Objective Prism, IIIaJ	PDS	$m_B \sim 18$	–
Hewett and Webster[3]	in progress	Schmidt Objective Prism, IIIaJ	APM	$m_B \sim 18.5$	~ 1000
Reboul *et al.*[4]	in progress	Berger & Fringant Catalogue[5]	eye	$m_B \sim 18.5$	~ 2000
Surdej *et al.*[6]	in progress	Véron-Cetty & Véron Catalogue[7]	various	$m_v < 18.5$ $M < -29$	~ 1000
Weedman and Djorgovski[8]	complete	CFHT grens	eye	$m_B < 20.5$	$\sim 200 - 400$

1. Djorgovski and Meylan (1988)
2. Hewitt and Burbidge (1987)
3. Webster *et al.* (1988a)
4. Reboul *et al.* (1987)
5. Berger and Fringant (1984)
6. Surdej *et al.* (1988)
7. Véron-Cetty and Véron (1987)
8. Weedman and Djorgovski (1988)

151 quasars, which gives $P_{miq}(\gtrsim 2\text{–}4'') \sim 1.3 \pm 0.9\%$. Reboul *et al.* (1987) are undertaking a systematic search among close pairs ($< 9''$) of blue images in the Berger-Fringant (1984) Catalogue. This group selected 62 pairs of objects for spectroscopic followup, and observations of 15 of these have produced no candidate multiply imaged quasars. In addition, Djorgovski and Meylan (1988) are undertaking high resolution observations of quasars selected from the Hewitt and Burbidge Catalogue (1987). The quasars are chosen to have $z > 1.5$ and $M_v < -28$, and $\sim$ 200–250 have been examined so far. One good lens candidate, with a separation of $\sim 6''$, one binary quasar and several cases of foreground galaxies superimposed on background quasars have been identified. Since there are other good candidates in the sample still to be checked, this gives $P_{miq}(\gtrsim 1'') \lesssim 0.4\%$. The APM Gravitational Lens Survey will be described in the next section.

In all these surveys, the base quasar catalogue and the number of detected lenses is small. In addition, different selection criteria have been used, so that the probabilities are not directly comparable. It is particularly difficult to assess how sensitive the surveys are to systems with a substantial magnitude difference between the components, and how well the small separation limit to image pairs is determined. Thus, the derived probabilities should be viewed with caution.

4. The APM Gravitational Lens Survey. The aim of the APM survey is to compile a statistical sample of gravitationally lensed quasars with $m_B \lesssim 18.75$. The Automated Plate Measuring machine in Cambridge is used to scan high quality UK Schmidt direct and objective-prism plates. Using the algorithms for quasar detection developed principally by Hewett (Hewett *et al.* 1985), a list of quasar candidates with $m_B \lesssim 18.75$ is defined. One of the key features of the survey is that all images, independent of morphology, are examined for evidence of quasar-like spectra. This procedure ensures that any multiply imaged quasars and quasar-galaxy associations are also identified in addition to the many "conventional" stellar-like quasar images. It is important to realize that in many quasar surveys to date, images that show evidence of structure, and particularly a nonstellar component, have been deliberately excluded from quasar candidate lists. By comparison, the quality of the UK Schmidt survey plates is such that faint galaxies and structure associated with the quasars can be seen some two magnitudes fainter than with POSS material. Following the automated selection procedures, the candidates are examined visually to exclude cases where the objects consist solely of a galaxy or group of galaxies. A number of such cases can not be reliably excluded by the automated procedures given the restricted parameterizations of the images that are available. Two groups of quasar candidates are obtained: the first has quasar-like spectra corresponding to direct images that show no evidence of image structure, the second has quasar-like spectra corresponding to images that show evidence of image structure. Both groups of objects are then observed spectroscopically at intermediate resolution; the first group as part of the Large Bright Quasar Survey (LBQS, Foltz *et al.* 1987), and the second as part of the automated search for lensed quasars. In addition, as part of the lens survey, all confirmed quasars with $z \geq 1$ from the LBQS are being imaged with $\lesssim 1''$ resolution to determine whether in fact any of these are multiple images with separations $\lesssim 2''$. From the sample for which observations have currently been completed, $P_{miq}(\theta \gtrsim 2.5'') \sim 0.3\%$.

The survey is extremely well defined, and the detection probability as a function of magnitude limit, component spatial separation, component magnitude difference, and quasar restframe spectrum is exactly specified. This allows the completeness of the sample to be determined. The technique can be used to survey efficiently large areas and has a proven quasar detection capability. In addition, the existence of a large sample of unlensed quasars, selected with identical techniques provides an invaluable control sample.

To date, a search for lensed quasars covering ~ 214 square degrees in nine UK Schmidt plates has been completed, including the spectroscopic followup. This sample contains 296 quasars with $16.5 \lesssim m_B \lesssim 18.75$. Spectroscopy of the control sample is considerably further advanced, with ~ 750 quasars spectroscopically confirmed. An MMT or Las Campanas spectrum with $S/N \sim 10$–15 at 4500Å with 6Å resolution is available for every quasar. The quasar sample is extremely effective over the redshift range 0.2–3.3. High resolution imaging of the sample will start in August 1988. The first results of the survey include detection of an excellent gravitational lens candidate, and statistical evidence for microlensing by foreground galaxies. This latter result emphasizes the importance of retaining images classified as galaxies in the lens candidate list. The gravitational lens candidate is 1429-0053 (Hewett *et al.* 1985), comprising two quasar images. The separation of the two images is 6.1″ and the magnitude difference is 2.9 m. The brighter image has an apparent magnitude of $m_B = 17.5$. The redshifts of the two images are 2.07, and the velocity difference between the two images is 260 ± 300 km s^{-1}. The brighter image shows two strong Mg II absorption line systems at redshifts 1.55 and 1.62. The quasars are not detected with the VLA at 2 cm wavelength at the 0.3 mJy level. High resolution (> 0.8″) CFHT CCD images show no additional flux between the images to $R \sim 25$. Figure 1 shows the spectra of the two stellar images and also the result of dividing the B spectrum by the A spectrum. There appear to be slight differences between the two spectra; however, it is difficult to determine whether this is due to different ratios of continuum-to-line strengths, or variation in the line profiles. The differences are no greater than the differences evident in other lensed quasars. Figure 2 shows the two images as they appear on the UK Schmidt plate.

The second result (Webster *et al.* 1988b) is that there is a large excess of quasars superimposed on galaxies that are visible on the Schmidt plates, compared to the number predicted under the assumption that there is no relation between the quasar and galaxy positions. The galaxy-quasar separations are 3–6″, and the galaxies have $m_B \lesssim 21$, consistent with the galaxies having redshifts $z \sim 0.1$–0.3. The total sample contains 296 quasars, 11 of which are superimposed on galaxies. Figure 2 shows examples of some of these objects. The quasar sample excludes objects with redshifts $z < 0.5$ to avoid confusion with galaxy and quasar

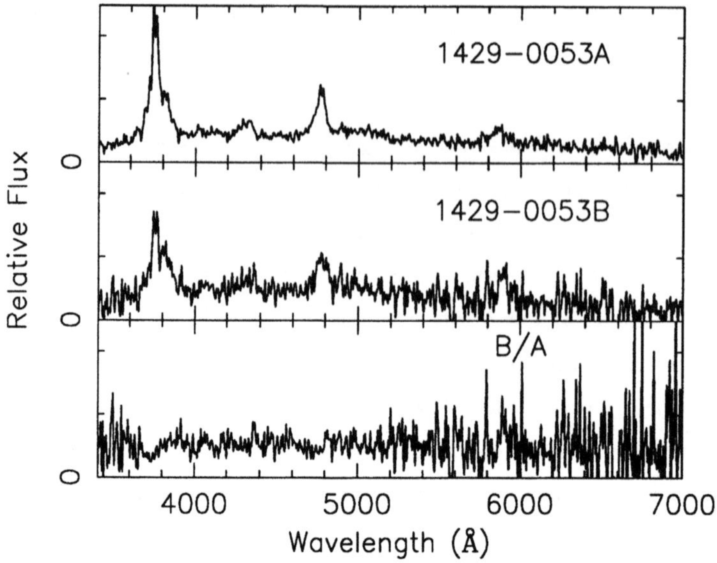

Fig. 1. The top two plots show the spectra of the two images of 1429-0053 from 3400–7000Å. The pixel size is 2.5Å, and each spectrum has been smoothed by a Gaussian function with a dispersion of one pixel. The third plot shows the B spectrum divided by the A spectrum, with an arbitrary rescaling.

pairs at the same redshift. The expected number of such associations is

$$\sim 2.6 \left(\frac{R}{6''}\right)^2 \left(\frac{N_g}{10^3}\right) \left(\frac{N_q}{296}\right)$$

where R is the separation in arcseconds, N_g is the surface density of galaxies per square degree to a limiting magnitude of 21, and N_q is the number of quasars in the sample. Thus, the probability of seeing 11 such instances is $\lesssim 0.01\%$. The most natural explanation of this result is that the quasars have been microlensed by compact objects associated with the galaxies. Both the redshift and absolute magnitude distributions of the "lensed" sample are consistent with those of the control sample. This is in accord with the predictions of microlensing. A possibly related observation described by Fugmann (1988), is 12 radio quasars that with $z > 1.7$ show an apparently significant excess of galaxies with $r < 23$ within a projected distance of one arcminute.

Our result suggests that microlensing may be particularly significant for high luminosity quasars since by using Schmidt plates we only search $\sim 30\%$ of the path length to the quasars. We are planning to obtain further information on these objects, and to triple the size of the sample to enable the effect to be quantified more rigorously.

5. Some Applications of Surveys. It is possible to use the results obtained in the previous section

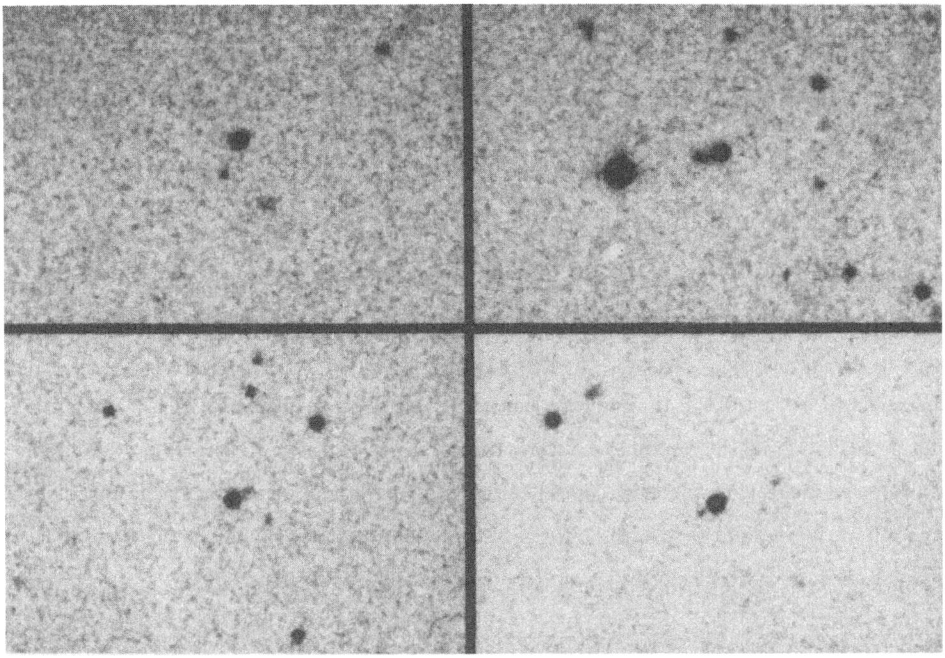

Fig. 2. Examples of quasar-galaxy associations taken from a single Schmidt field. In each case the quasar is in the center of the field. In the upper left, the galaxy is the faint extension on the upper right of the quasar image. This quasar has a redshift of 0.8, and the separation of the images is $\sim 3''$. The quasar in the upper right has a redshift of 2.1 and the separation is $\sim 8''$. This quasar was not included in the statistical sample. The quasar in the lower left has a redshift of 1.4 and a separation of $\sim 6''$. The final example in the lower right is the lens candidate, 1429-0053.

to estimate the amount of mass associated with the galaxies visible on the Schmidt plates. Firstly, assume that the lens is an isothermal mass distribution comprising compact objects, each individually able to amplify the quasar. The observed overdensity of quasars near galaxies can then be used to estimate the velocity dispersion σ_v in the lens. Using the terminology and calculations of Schneider (1987), we can write the fractional number enhancement as

$$\delta N/N \sim 0.06\, f(L) \left(\frac{\sigma_v}{250 \text{ km s}^{-1}}\right)^2 \left(\frac{\theta}{1'}\right)^{-1} ,$$

where θ is the area searched around each galaxy. The constant $f(L)$ gives a measure of the bias introduced by lensing, and a conservative estimate from Schneider gives $f(L) < 1.7$. Thus,

$$\sigma_v \gtrsim 450 \text{ km s}^{-1} \left(\frac{\theta}{6''}\right)^{0.5} \left(\frac{1.7}{f(L)}\right)^{0.5} ,$$

where the Poisson error in the estimate is ~ 40 km s^{-1}. It is unlikely that a second image of the quasar exists at an equal or greater distance on the opposite side of the galaxy. Any such image, which was only a few magnitudes fainter than the primary image, would be visible on the Schmidt plates. If microlensing does enhance the number of quasars we see, then there must be more mass associated with these galaxies than is normally expected for single galaxies and their halos.

An alternative model is that the quasars are amplified, but singly imaged. Then, for a mass distribution with a power-law form for the surface density, $\Sigma = \Sigma_0 h^k$, where Σ_0 is constant, h is the image impact parameter and $0 > k > -2$. The amplification is $[(1-x)(1-(k+1)x)]^{-1}$ where $x = 8\pi\mathcal{D}(G/c^2)\Sigma_0 h^k/(k+2)$. For our data, the mean amplification is ~ 1.8, and thus for a constant density mass distribution ($k = 0$), $\Sigma_0 \sim 4 \times 10^2\,(10^3\,Mpc/\mathcal{D})\;M_\odot\;\mathrm{pc}^{-2}$. Such lenses would be more effective at lower redshifts. Thus, if the tidal effects of the mass concentration can be ignored, an average surface density of this magnitude is required to give the observed effect. This value is similar to the surface density of the cores of dark matter halos of lower-luminosity galaxies (Kormendy 1988).

For all the surveys that are currently being undertaken, numbers are still small and therefore subject to large errors. However, an adequate sample of gravitational lenses would provide the data to determine both the bias and the normalization due to Ω_{lens}. In addition, it is also possible to determine the shape of the mass function for clumps with $M \gtrsim 10^{11}\,M_\odot$ and $\Sigma \gtrsim \Sigma_{crit}$. Using the data from Surdej *et al.* (1988) and the APM Survey (these have comparable selection criteria for this calculation), one might estimate $P_{miq}(m \lesssim 18.5, \theta \sim 2.5'') \sim 10^{-2.5}$ and $P_{miq}(m \lesssim 18.5, \theta \sim 1'') \sim 10^{-1.7}$. Since $P_{miq} \sim \Omega(> M)$ and $M \sim \theta^2$, then $N(> M) \sim M^{-2}$, where $N(> M)$ is the number of objects with mass greater than M. As the data sets improve, it will be possible to determine this function in more detail.

One of the uncertainties that the surveys for gravitational lenses may be able to answer is whether or not the wide separation lens candidates are lenses or binary galaxies. There are now three cases where the two spectra are remarkably similar, and yet there is no evidence of light from a lensing object. If these objects are lensed, then the mass-to-light ratio of the lens is many thousands, and the objects are more massive than galaxies. Gravitational lensing will provide a method for detecting such mass concentrations, if they exist.

References

Berger, J. and Fringant, A. M. 1984, *Astron. Astrophys. Suppl.*, **58**, 565.

Crampton, D., Cowley, A. P., Hickson P., Kindl, E., Wagner, R. M., Tyson, J. A. and Gullixson, C. 1988, *Astrophys. J.*, **330**, 184.

Djorgovski, S. G., Perley, R., Meylan, G. and McCarthy, P. 1987, *Astrophys. J. Lett.*, **321**, L17.

Djorgovski, S. G. and Meylan, G. 1988, this volume.

Dyer, C. C. and Oattes, L.M. 1988, *Astrophys. J.*, **326**, 50.

Foltz, C. B., Chaffee, F. H., Hewett, P.C., MacAlpine, G. M., Turnshek, D. A., Weymann, R. J. and Anderson, S. I. 1987, *Astron. J.*, **94**, 1423.

Foltz, C. B. and Osmer, P. S. 1988, A.S.P. Conference Series 2, edited by P. S. Osmer, A. C. Porter, R. F. Green, and C. B. Foltz (Brigham Young: Utah), p. 361.

Fugmann, W. 1988, *Astron. Astrophys.*, , in press.

Gaston, B. 1983, *Astrophys. J.*, **272**, 411.

Hewett, P. C., Irwin, M. J., Bunclark, P., Bridgeland, M. J. and Kibblewhite, E. J. 1985, *Mon. Not. R. astr. Soc.*, **213**, 971.

Hewitt, A. and Burbidge, G. 1987, *Astrophys. J. Suppl. Ser.*, **63**, 1.

Hewitt, J. N., Turner, E. L., Schneider, D. P., Burke, B. F., Langston, G. I. and Lawrence, C. R. 1988a, *Nature*, **333**, 537.

Hewitt, J. N., *et al.* 1988b, this volume.

Kayser, R., Refsdal, S., and Stabell, R. 1986, *Astron. Astrophys.*, **166**, 36.

Kormendy, J. 1988, Proc. First Yellow Mountain Summer School on Physics and Astrophysics, in press.

Lavery, R. J. and Henry, J. P. 1988, *Astrophys. J. Lett.*, **329**, L21.

Lynds, R. and Petrosian, V. 1986, *Bull.A.A.S.*, **18**, 1014.

Maccacaro, T., Gioia, I. and Stocke, J. 1984, *Astrophys. J.*, **283**, 486.

Phinney, E. S. and Blandford, R. D. 1986, *Nature*, **321**, 569.

Reboul, H., Vanderreist, C., Fringant, A. M. and Cayrel, R. 1987, *Astron. Astrophys.*, **177**, 337.

Schneider, P. 1987, *Astron. Astrophys.*, **179**, 80.

Shanks, T., Fong, R., Boyle, B. J. and Peterson, B. A. 1987, *Mon. Not. R. astr. Soc.*, **227**, 739.

Soucail, G., Fort, B., Mellier, Y. and Picat, J. P. 1987, *Astron. Astrophys.*, **172**, L14.

Sramek, R. A. and Weedman, D. W. 1978, *Astrophys. J.*, **221**, 468.

Surdej, J., Swings, J. -P., Magain, P., Borgeest, U., Kayser, R., Refsdal, S., Courvoisier, T. J.-L., Kellermann, K. I. and Kuhr, H. 1988, A.S.P. Conference Series 2, edited by P. S. Osmer, A. C. Porter, R. F. Green, and C. B. Foltz (Brigham Young: Utah), p. 183..

Turner, E. L. 1980, *Astrophys. J. Lett.*, **242**, L135.

Turner, E. L., Ostriker, J. P. and Gott, J. R. 1984, *Astrophys. J.*, **284**, 1.

Véron-Cetty, M. -P. and Véron, P. 1987, ESO Scientific Report, Number 5.

Walsh, D., Carswell, R. F. and Weymann, R. J. 1979, *Nature*, **279**, 381.

Webster, R. L., Hewett, P. C. and Irwin, M. J. 1988a, *Astron. J.*, **95**, 19.

Webster, R. L., Hewett, P. C., Harding, M. E. and Wegner, G. A. 1988b, preprint.

Weedman, D. and Djorgovski S. G. 1988, preprint.

Zwicky, F. 1937, *Phys. Rev.*, **51**, 679.

AN OPTICAL IMAGING SURVEY FOR GRAVITATIONAL LENSES AND THE DISCOVERY OF A NEW LENS CANDIDATE

S. Djorgovski
Palomar Observatory, California Institute of Technology
Pasadena, CA 91125, USA

G. Meylan
European Southern Observatory
D-8046 Garching bei München, FRG

Abstract. We are conducting an optical imaging search for gravitational lenses. A sample of known quasars was selected on the basis of high redshifts and apparently large absolute luminosities; these selection criteria significantly enhance the probability that a given QSO is gravitationally magnified. So far, our search has yielded at least one new gravitational lens system, associated with the quasar UM 425 = 1120+019, one binary quasar, PKS 1145–071, and several other promising candidates that still need spectroscopic confirmation. In addition, we found several cases of foreground galaxies within a few arcsec from the quasars, which should be useful for absorption line studies. The optical image of the UM 425 system consists of four or more components, at least two of which have identical spectra. There are many faint galaxies in the field, suggestive of a lensing cluster at a redshift of several tenths, and tentative evidence for a $z \simeq 0.6$ galaxy underlying the second-brightest quasar image. We conclude that the heuristically designed optical searches for gravitational lenses can successfully complement the radio-based surveys.

The survey. From the extensive modeling of Turner, Ostriker and Gott (1984), Ostriker and Vietri (1986), and Blandford and Kochanek (1987), we know that the distribution of lensed image separations should peak near 0.5 arcsec, so that the frequency of lensed QSO's increases rapidly with the resolution limit. This condition is even more pronounced in a flux-limited sample (i.e., virtually all QSO samples), because of the magnification biasing. It is thus profitable to preselect the QSO's for a lens search by their apparent absolute luminosity: since the luminosity function falls off steeply at every redshift, an apparently high-luminosity QSO has a higher a *priori* probability of being lensed. One can improve the chances even further by selecting the high-redshift QSO's because they have large intercept lengths. A QSO sample preselected in this way should have a considerably higher lens frequency than a "random" QSO catalog. The exact probability enhancement is difficult to compute, since the biasing

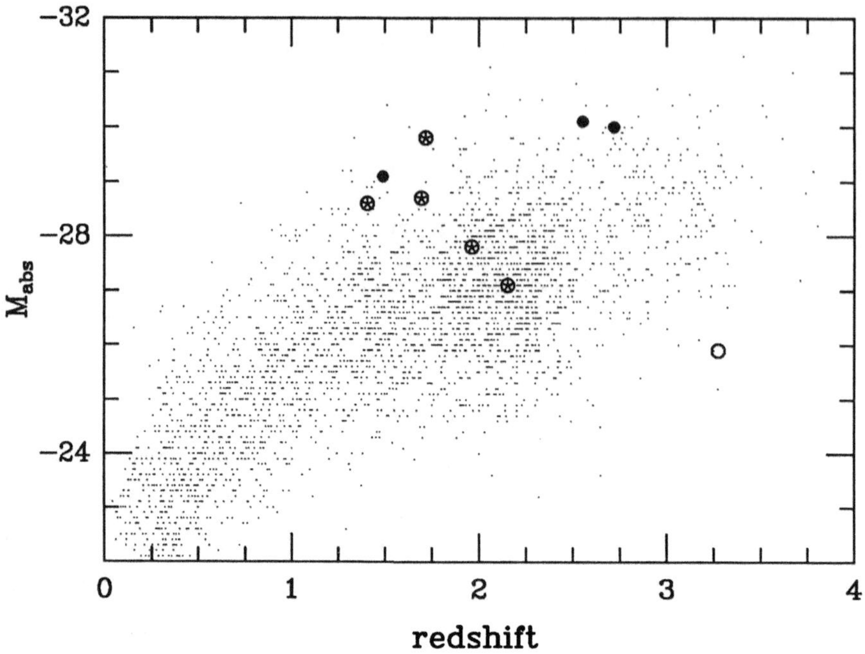

Fig. 1. The bivariate distribution of redshifts and absolute magnitudes of QSO's from Hewitt and Burbidge (1987). The solid circles denote the optically selected lenses, as described above; the star-crossed circles denote the optically selected, but serendipitously discovered, lenses; and the open circle denotes the one radio selected lens (other radio selected lenses have no measured redshifts as yet, but they are also very faint). The effects of magnification biasing for the optically selected lenses are apparent.

function depends on many unknowns (luminosity function of QSO's at different redshifts, mass distribution and clustering on large scales, cosmology, etc.). Figure 1 shows the effect of redshift and luminosity selection on the sample of quasars from the catalog of Hewitt and Burbidge (1987).

Most QSO's are identified on a Schmidt plate material (e.g., the sky surveys) and thus with a resolution limit $\sim$ 2–5 arcsec, depending on the seeing, magnitude contrast, etc. Thus, good-seeing CCD imaging of a well-selected sample of QSO's previously detected on a Schmidt plate material may readily produce some good lens candidates, with $\sim$ 1 arcsec or larger separations. Discrepant colors can be used to eliminate spurious associations with foreground stars. Using a variety of image processing methods, one can push the detection of lensed pairs or multiples well into the seeing limit. Any promising candidates are then checked spectroscopically.

We have thus conducted for the last few years an optical (CCD) imaging search for lensed quasars from several different observatories. A sample of high-redshift QSO's with apparently high absolute luminosities has been selected from the Hewitt and Burbidge (1987) catalog. Most data were obtained as a byproduct of other projects: all of our "blank sky" frames (needed for CCD calibration) always have a carefully chosen QSO in them. So far, we have one very good lens candidate, UM 425, described below, one probable binary quasar, PKS 1145–071 (Djorgovski *et al.* 1987), and several other promising candidates. In addition, there are several cases of QSO's with foreground galaxies within a few arcsec, which should be useful for absorption-line studies, or monitoring for gravitational microlensing. Our results will be described in more detail in future papers. An equivalent, very successful survey has been conducted independently by the Liège group (Surdej *et al.* 1987; Magain *et al.* 1988).

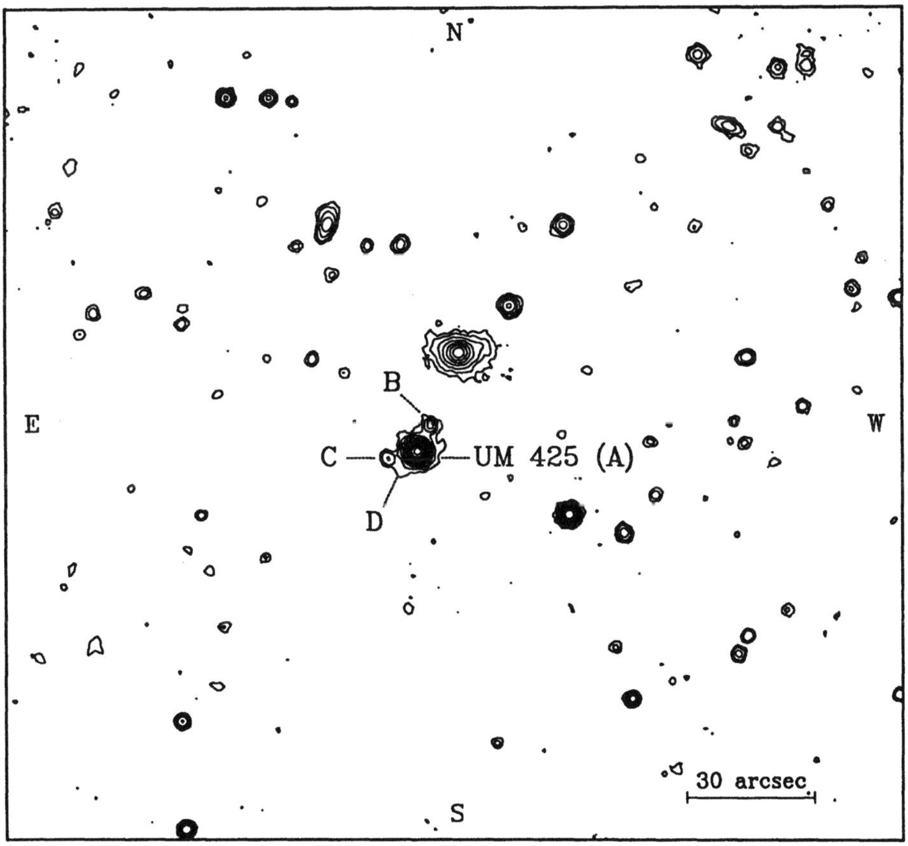

Fig. 2. A section of a $V + R$ stack image of the UM 425 field, obtained at CTIO. The field shown is 190×212 arcsec, with north to the top, east to the left. The components of the UM 425 system are indicated with the letters. The isophotal levels are spaced logarithmically in factors of 2.

UM 425: A New Lens Candidate. Multicolor CCD images of the object, obtained at CTIO in March of 1987 (Fig. 2) show several close companions (within 3 to 6 arcsec), whose *BVR* colors appear to be similar or identical to those of the principal quasar image (UM 425 A) itself. There appears to be a moderately distant ($z \sim 0.5$?) foreground cluster in the field. The apparent magnitude of the component A is $V \simeq 16.2$; the components B and C are $\sim 4.6^m$ and 5.6^m dimmer in *V*, respectively. The component B has the same *BVR* colors as the component A, within the measurement errors; the colors of C are slightly different. We cannot say much about the fainter, close component D, other than it is there, in all of our CCD images. Both components C and D appear slightly nonstellar. The brightest component, A, appears unresolved on our CCD images (seeing FWHM $\sim$ 1.5 arcsec). R. Perley kindly obtained a brief VLA snapshot of the field and established that the system is not brighter than about 0.3 mJy at 6 and 20 cm.

The spectra of the components A and B show the same emission lines, as shown in Figure 3. The redshift of the quasar, based on the "clean" C III] 1909 line is $z = 1.465 \pm 0.005$.

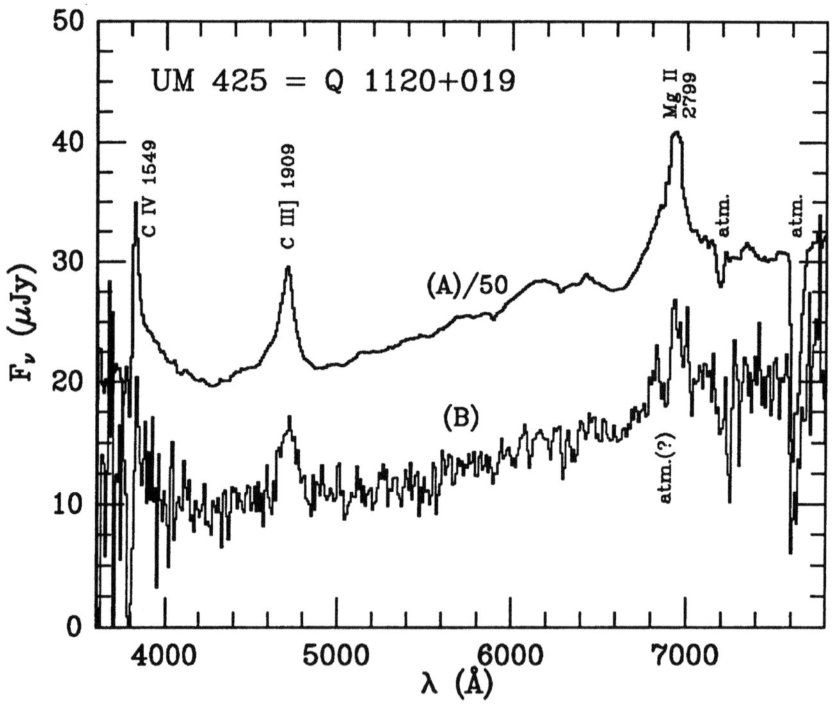

Fig. 3. CCD spectra of components A and B, obtained at ESO. The brighter component (A) was scaled down by a factor of 50, for an easier comparison. The two spectra coincide in shape within the measurement noise, and the equivalent widths of the emission lines are the same, within the error bars. The redshift difference is consistent with zero to within $\sim$ 200 km s^{-1}.

The cross-correlation redshift difference between the components A and B is consistent with zero, with ~ 200 km s^{-1} uncertainty. The relatively bright galaxy NNW from the quasar is at $z = 0.1265$ and is probably unrelated to the system. The evidence in favor of the gravitational lens interpretation comes from the similarity of observed spectra, both in the shape, and equivalent widths of emission lines. There may be some associated absorption in the Mg II 2799 and in C IV 1549 lines, but the data are affected by the noise shortward of 3800Å (C IV) and the atmospheric absorption (Mg II), which make the interpretation difficult. Overall, the gravitational lens interpretation of the data seems most natural to us.

Prompted by the slightly redder color of the component B, we scaled the spectrum of the component A by 0.01 and subtracted it from the spectrum of the component B. The residual can be interpreted as a spectrum of a $R \simeq 22.5$ mag galaxy at $z \simeq 0.6$ (Fig. 4). This is consistent with our guesstimate for the redshift of the tentative foreground cluster in this field, and it would correspond to a favorable geometry for gravitational lensing.

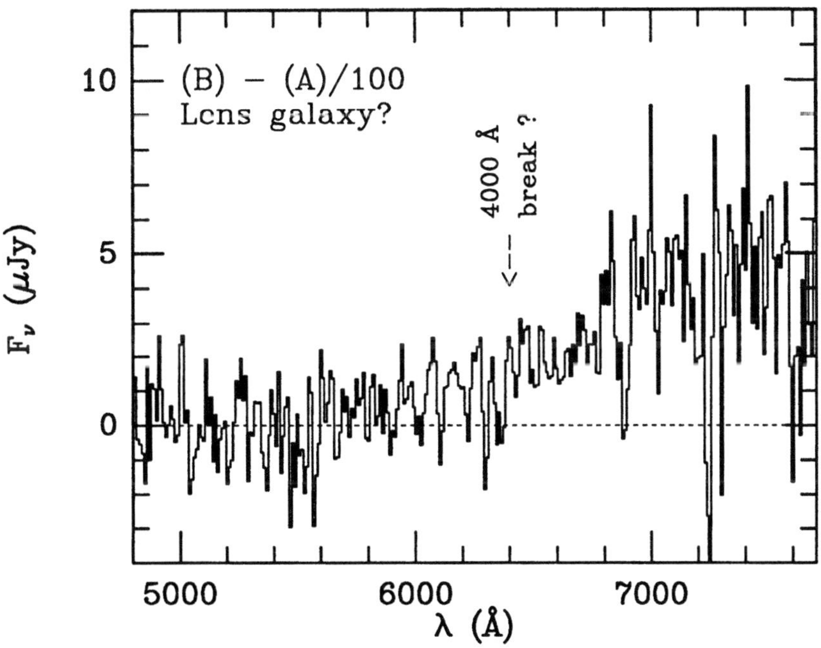

Fig. 4. Residual spectrum obtained after dividing the spectrum of the component A by 100, and subtracting it from the spectrum of the component B. The difference, shown here, is reminiscent of an early-type galaxy spectrum at a redshift $z \simeq 0.6$, if the continuum rise at ~ 6400Å is attributed to the 4000Å break.

Conclusions. The results so far are encouraging. Systematic optical searches for gravitational lenses, using the redshift and luminosity as the selection criteria (e.g., by us, and the Liège group) have led to the discovery of three new gravitational lenses, and more should be forthcoming. The yield is at least comparable to other lens surveys (e.g., using the radio-selected samples), and it was achieved with a moderate expenditure of the telescope time. Moreover, we are likely missing some narrow-separation lenses in our average-seeing ground-based imaging, and it may be worthwhile to repeat the observations of the most luminous high-redshift quasars, from a good-seeing site (e.g., Mauna Kea), with some seeing-compensating technique, or with the Hubble Space Telescope.

We are greatly indebted to the staffs of CTIO, ESO, KPNO, MMTO, Palomar, and Las Campanas observatories for their valuable and professional help during the many observing runs. Rick Perley, Peter Shaver, and others, collaborated with us on several projects related to our survey. Finally, we acknowledge partial financial support from California Institute of Technology (S.D.) and European Southern Observatory (G.M.).

References.

Blandford, R., and Kochanek, C. 1987, *Ap. J.*, **321**, 658.

Djorgovski, S., Perley, R., Meylan, G., and McCarthy, P. 1987, *Ap. J. (Letters)*, **321**, L1.

Hewitt, A., and Burbidge, G. 1987, *Ap. J. Suppl.*, **63**, 1.

Magain, P., Surdej, J., Swings, J.-P., Borgeest, U., Kayser, R., Kühr, H., Refsdal, S., and Rémy, M. 1988, *Nature*, **334**, 325.

Ostriker, J., and Vietri, M. 1986, *Ap. J.*, **300**, 68.

Surdej, J., Magain, P., Swings, J.-P., Borgeest, U., Courvoisier, T.J.-L., Kayser, R., Kellermann, K.I., Kühr, H., and Refsdal, S. 1987, *Nature*, **329**, 695.

Turner, E., Ostriker, J., and Gott, J.R. 1984, *Ap. J.*, **284**, 1.

STATISTICS OF GRAVITATIONAL LENSES: GALAXIES AND DARK MATTER

Lawrence M. Krauss[1]
Center for Theoretical Physics and Dept. of Astronomy
Sloane Laboratory, Yale University
New Haven CT 06511

Introduction. The gravitational lensing of distant objects (Zwicky,1937) has become a subject of intense current interest with the discovery over the last 7 years of at least as many multiply imaged quasar candidates (see Burke, 1984 or Hewitt et al., 1988). Indeed, as I write this, I have just read of yet another 4 image candidates. In most cases of quasar lensing, image splittings are generally on the order of a few arc seconds. As lensing statistics improve (i.e. as more lenses are discovered), it becomes worthwhile to ask whether the observed lensing characteristics are in accord with a priori expectations. However, in order to do this one must have some a priori expectations! This requires modelling both the quasar distribution as well as the distribution of intervening lensing objects, which include stars, galaxies, clusters of galaxies, and unknown objects. Several authors (Turner, Ostriker, and Gott 1984, herein referred to as TOG; Dyer 1984; Anderson and Alcock 1986; Ostriker and Vietri 1986) have investigated what expectations one might have for the distribution of angular image separations, based on the known galaxy and cluster mass distributions. In addition, much work has been done to understand the effects of image amplification on a complete, magnitude limited sample of quasars (Turner 1980; Avni 1981; Tyson 1981; Vietri and Ostriker 1983; Vietri 1985). Also, the lensing characteristics of other, more exotic, systems such as cosmic strings have recently been analyzed (Vilenkin, 1984, Gott 1985). Finally, it has long been recognized that various general cosmological parameters such as the mean mass density or the Hubble constant may also be probed by lens observations (see Turner, 1988).

Up to this point, the standard analysis has been that of TOG, who examined the consequences of assuming either a point mass distribution, or, more plausibly, a distribution of galactic systems each of which was represented by an isothermal sphere. Dyer on the other hand, had numerically analyzed phenomenological models in which galaxies were assumed to have either a King or de Vaucouleurs mass distribution. His results on the distribution of angular image separations differed from those of TOG, and provided perhaps the first indication that a finite core radius might play an important role.

[1] Also Visiting Scientist, Harvard-Smithsonian Center for Astrophysics. Research supported in part by a Presidential Young Investigator Award and by the DOE

However, since his methods were primarily numerically based, and since the asymptotic behavior of the distributions he examined were different, it is difficult to ascribe any particular root for the difference between his results and TOG. In addition, it is worthwhile to continue to focus on a distribution that is at least asymptotically isothermal (i.e. falls as $1/r^2$) if one wants to seriously take into account the effects of dark matter distributions.

This early work suggested some potential problems when theory and observation are compared. In the first place, a number of the lens candidates have no observable associated lensing objects. In addition, the mean splitting of images is slightly larger than predicted by TOG. Of course, the present sample is very small, and selection effects can be significant. Nevertheless, there may be more out there than meets the eye.

With these things in mind, Gary Hinshaw, with minor assistance from myself, extended the work of TOG on lensing by isothermal spheres by replacing the singularity at the origin by a finite core (Hinshaw and Krauss, 1987). I will summarize here the major results that appeared in our 1987 paper, and also mention a few results that have been derived since (see Hinshaw 1988). The simple analytical form of the distribution we used made it possible to derive rather general results. To our surprise, even a small core radius can have dramatic effects on image separations, lensing probabilities and other lensing characteristics. Galactic lensing may be rarer than expected, and resultant mean splittings and amplification of quasar images can be altered compared to the case of isothermal spheres. In particular we find that the mean image separation due to galaxies can be increased to agree better with present observations. We believe this simple analytical formalism can be used successfully to derive realistic predictions for the lensing properties of galactic systems. The only feature that has not yet been included is the contribution of the disk for spiral galaxies, which we feel will not qualitatively alter our predictions.

The effects we investigated are of course most significant for systems with the largest core radii. Among these might be possible "failed galaxies" consisting predominantly of dark matter. Such objects are favored, in order to reconcile the theoretically attractive notion of a flat universe with observation. It has recently been proposed that galaxies might not trace the dominant mass in the universe (for a review, see Peebles, 1986). Rather, in such biased galaxy formation models, only regions with unusually large primordial overdensities form galaxies, while in most regions galaxy formation is inhibited. In undertaking our work, we have been particularly interested in the role "failed galaxies" might play in altering lensing statistics. We find that the core radius effect suppresses any role they might have in producing multiple images, or in altering the imaging due to visible galaxies. However, they can significantly alter the observed quasar luminosity function without producing double imaging. Our formalism provides a useful and realistic framework not just for examining these questions, but also possiblly for calculating the statistics of Einstein Ring-like objects that have recently been sighted. These and other lensing occurences may then provide important information not just on the distribution of visible matter in the universe, but also on questions as diverse as the nature of dark matter, the origin of large scale structure, and the luminosity function of distant quasars.

The Model. In order to smooth out the unphysical singularity at the origin of an isothermal sphere, we considered a lens that has a density profile of the form,

$$\rho(r)= \sigma_{\|}{}^2/\{2\pi G[r^2+r_c{}^2]\} \quad (1)$$

where $\sigma_{\|}$ is the one-component velocity dispersion and r_c is the core radius (i.e. a plot of M(r)/r turns over at $r \approx r_c$). The projected surface density then has the form,

$$\Sigma(a)= \pi\sigma_{\|}{}^2/\{2\pi G[a^2+r_c{}^2]^{1/2}\} \quad (2)$$

where a is the radius transverse to the line of sight. The projected mass within a radius a is given by,

$$M(a)= \pi\sigma_{\|}{}^2\{[a^2+r_c{}^2]^{1/2} -r_c\}/G \quad (3)$$

Using standard techniques (Young et al 1980) the bending angle of light passing through the lens as a function of impact parameter b is then found to be,

$$\alpha(b)= \alpha_0[(b^2+r_c{}^2)^{1/2} -r_c]/b \quad (4)$$

where $\alpha_0= 4\pi\sigma_{\|}{}^2/c^2$ is the (constant) bending angle for a corresponding singular isothermal sphere (obtained by letting $r_c \to 0$).

The ray trajectories for this lens model yield an equation for the impact parameter b in terms of the core radius and the transverse distance l, of the lens center from the line of sight, shown below:

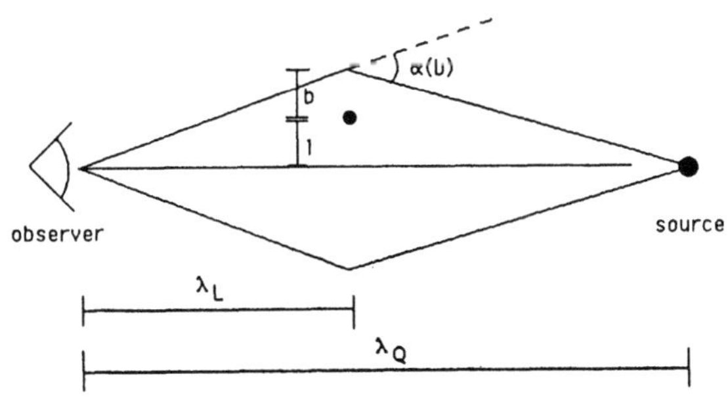

One obtains

$$b+l= a_{cr}[(b^2+r_c{}^2)^{1/2}-r_c]/b \quad (5)$$

where a_{cr} is the critical (i.e. largest) radius for multiple imaging by the corresponding singular lens (TOG; Hinshaw and Krauss, 1987). It is this radius that sets the fundamental length scale against which the core radius should be compared.

If one plots both sides of (5) as a Young diagram (see Hinshaw and Krauss, 1987) for various values of the parameter $\beta=r_c/a_{cr}$, it is clear that there exists a critical value of β, below which there exist three solutions for b, and above which there is only a single solution. By equating slopes at the origin the critical value is found to be $\beta_{cr}=1/2$. This implies that multiple image formation is only possible if the core radius satisfies,

$$r_c < a_{cr}/2 \qquad (6)$$

It is important to emphasize that a_{cr}, the multiple imaging radius for a singular sphere, is a function of both the lens and source positions (as well as of $\sigma_{||}{}^2$). Generally a_{cr} is maximal for a lens roughly midway between source and observer, while it goes to zero as the lens approaches the source or the observer. It follows that for a given core radius (neglecting lens evolution), there will always be a region of lens redshift space where multiple image formation is impossible, no matter how close the lens is to the unperturbed line of sight (see Hinshaw and Krauss, 1987; Hinshaw 1988).

A detailed study of the solutions of (5) yields the following formula for l_0, the radius of the disc (centered on the unperturbed line of sight) in which the lens must lie in order to produce multiple images. l_0 thus plays the role of a_{cr} for a finite core lens. We find (see Hinshaw, 1988 for further details)

$$l_0{}^2=[a_{cr}{}^2+5a_{cr}r_c-r_c{}^2/2]-0.5r_c{}^{1/2}(r_c+4\,a_{cr}) \qquad (7)$$

We then finally define the cross section for multiple image formation as $\sigma_{sc}=\pi l_0{}^2$, which gives

$$\sigma_{sc}(\beta)=\pi a_{cr}{}^2[(1+5\beta-\beta^2/2)-0.5\beta^{1/2}(\beta+4)^{3/2}] \qquad \beta<1/2 \qquad (8)$$

$$\sigma_{sc}(\beta)=0 \qquad \beta>1/2 \qquad (9)$$

Note the rather severe dependence of σ on β; for example, at $\beta=0.15$ the cross section is diminished by a factor of ten relative to the singular lens cross section. This sensitivity indicates that lenses with finite core radii can have significantly different statistics than their singular counterparts. This is one of our chief results.

Next, at least for the case when the lens is on the line of sight (l=0) one can show analytically that the angular separation of the two outer images is reduced by a factor $(1-2\beta)^{1/2}$ relative to the images produced by a singular sphere. A study of the full solutions reveals that this separation is only weakly dependent on l (Hinshaw, 1988). Hence while lensing cross sections can be severely reduced, image separation is relatively insensitive to the core radius. Since angular splitting falls off much less rapidly with β than do cross sections, this suggests that small angular splittings can be suppressed.

Finally, the focussing of light by a gravitational lens will generally result in a magnification of quasar images, independent of whether multiple images are formed. This magnification can be extremely large for small impact parameters (see Hinshaw and Krauss, 1987). Moreover, significant image magnification can occur even when β is greater than 1/2, so that there is only one image.

Hinshaw (1988) has also analytically computed differential and integrated optical depths (i.e. probabilities) for lensing as a function of source and lens redshift. Since cross sections are smaller for finite core radii, the optical depth is in general severely reduced for the case of non-zero compared to zero β. Most interesting, perhaps, because multiple image formation is only possible if the core radius is less than half the critical radius for lensing, and because the critical radius shrinks as the source moves closer to the observer, there exists a critical source redshift below which multiple image formation is impossible. In particular Hinshaw finds that multiple image formation will not be possible for a lens or source *at any redshift* (for an $\Omega=1$ universe) unless:

$$r_c/\alpha_0 R_0 < .07, \quad \text{where } \alpha_0 R_0 = 37.8\ h^{-1}\ \text{kpc}\ (\sigma/300\ \text{km/sec})^2. \tag{10}$$

The Results:(a) Visible Galaxies. I now interpret the equations of the previous section using the numbers appropriate for galactic systems. The bending angle α_0 for an isothermal sphere, given in equation (4), has a numerical value of 1.26×10^{-5} (σ/300 km/sec)^2 radians [=2.59" (σ/300 km/sec)^2]. Similarly, from eq. (7) one finds, setting the Hubble radius, R_0 equal to 3000 h^{-1} Mpc (where H_0=100h km sec.$^{-1}$Mpc^{-1}), that $a_{cr} \leq O[5h^{-1}\ (\sigma/300\ \text{km/sec})^2]$ Kpc for a quasar redshift of order 1.

For a typical elliptical galaxy (characterized by $\sigma_{\|}\approx306$ km/sec for an L* galaxy, as defined in equation 12 below), the critical radius a_{cr} is less than, or on the order of, 2-6 Kpc. Then if the galactic core radius is on the order of 0.2 Kpc or greater, one would expect the statistical parameters of lensing by galaxies to be affected by our results, in certain cases by as much as an order of magnitude (Hinshaw and Krauss 1987; Hinshaw 1988). The observed rotation curves of spiral galaxies suggest that they have core radii of several Kpc (e.g. see Burstein and Rubin 1985). Elliptical galaxies probably have smaller core radii, but these radii are still likely to be in the range mentioned above for lensing statistics to be affected.

Specifically, to investigate whether the shape of the distribution of expected lensing events can be altered we compute the probability distribution for angular splittings $\Delta\theta$, $dP(\Delta\theta, Z_L \mid Z_Q)$, defined by

$$dP(\Delta\theta, Z_L \mid Z_Q) = \sigma(\Delta\theta)[\Sigma n_i(\Delta\theta,Z_L)d(\Delta\theta)]c\,dt(Z_L) \tag{11}$$

where $n_i(\Delta\theta,Z_L)$ is the number density of galactic lenses of type i (i.e elliptical, spiral, etc.) with the appropriate properties to produce a splitting of $\Delta\theta$. With some effort this can be determined once we choose a galactic luminosity distribution function that, to facilitate comparison with TOG, we take to have the Schechter (1976) form,

$$\Phi(L)dL = \phi^*(L/L_*)^\alpha\, e^{-(L/L_*)}\, d(L/L_*) \tag{12}$$

(where ϕ^* is a number density whose normalization will not be important here, and L_* and α are parameters that are fit to observation. In addition we assume $L/L^*=\sigma_{\|}/\sigma_{\|}{}^*$).

The cross section, $\sigma(\Delta\theta)$, is a complicated function of r_c (see Hinshaw and Krauss 1987). However, it can be shown that for a given core radius, r_c, the cross section to produce a splitting $\Delta\theta$ is substantially more suppressed, relative to the singular case, for small splittings than for large ones. This clearly implies that the distribution $P(\Delta\theta)$ will be skewed towards larger splittings compared to its form in the case of singular lenses. For the purposes of this discussion we assume that r_c remains constant as a function of luminosity, which is a reasonable approximation, at least for spiral galaxies. In this case we plot below $P(\Delta\theta)$ for elliptical galaxies (characterized by $\sigma_{\|}{}^{*}$ =306 km/sec in eq. (12)) which results from integrating the distribution in eq. (11) over all intermediate lens redshifts, for the particular case r_c=0.2 kpc. (Note that all curves have been normalized such that the area under each is unity.) Even in this case, the mean splitting is 50% larger than the singular case.

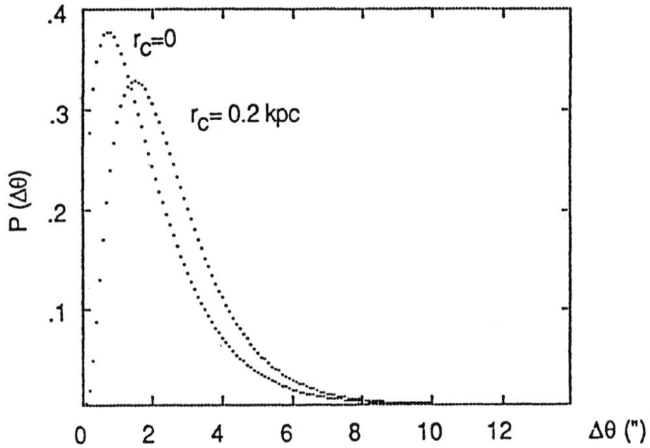

These results are quite striking and suggest that the large splittings observed in the 7 or so known lenses may not be out of line with the statistical expectations for lensing by galactic systems. Indeed, just for fun I plot below a renormalized version of the figure above, where I have superimposed a histogram containing 7 lens candidate observations compiled by Hewitt (1986). Ignoring the arbitrary normalization of the curve, one can see that the form of the finite core prediction agrees much better with this very preliminary data then does the singular lens prediction.

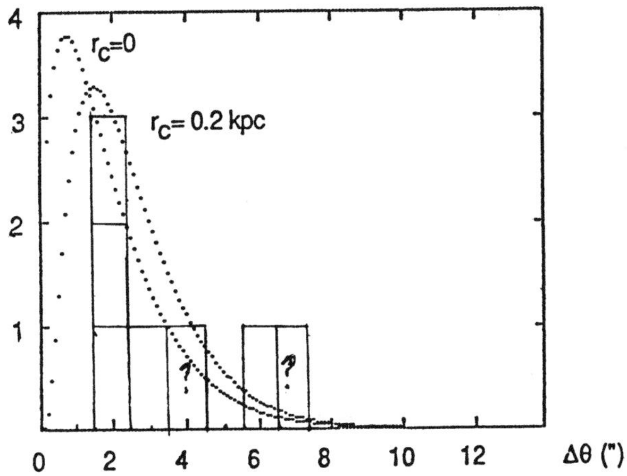

To summarize our results thus far: (1) Finite core effects can suppress the total optical depth for lensing by a factor of from 10 to10^5. If future observations suggest a lensing frequency that is in accord with the results of TOG we may be required to postulate the existence of non-galactic lenses; (2) Lensing by spiral galaxies can be drastically reduced compared to lensing by ellipticals, because of the larger core sizes and smaller central velocity dispersions of spirals. We find that for a mean core radii of 1 Kpc, and $\sigma_{||} =$ 177 km/sec for an L* galaxy, lensing by spirals is a factor of 6×10^{-5} less probable than by ellipticals (with assumed core radii of 0.2 Kpc), even though spirals make up about 70% of all galaxies by number. Indeed, eq. (10) provides a lower limit on velocity dispersions for spirals so that multiple image formation is possible, with the assumption of a core radius in excess of, say, 1 kpc; (3) The distribution of galactic lens redshifts is additionally suppressed for both small and large redshifts, due to the relative decrease at these redshifts in critical radii compared to the core radii. The redshift at which the distribution has its maximum is also reduced (Hinshaw, 1988); (4) Lensing of quasars at low redshifts should be additionally suppressed for a similar reason.

Of course, the actual probability distribution for angular splittings due to galactic lensing events can differ from our calculations due to several different theoretical and observational factors. First, our model for galaxies, while it is more refined than the simple isothermal sphere approximation, is still idealized. Next, the simple assumption that r_c is constant as a function of galactic size is an over-simplification, which can be remedied by considering a variety of models in which r_c is weakly correlated with galactic luminosity. Neither of these complications is expected to qualitatively affect our results.

One factor that could severely affect these predictions, however, is the bias introduced by observational selection effects. In particular, gravitationally lensed quasars will be preferentially selected in a flux limited sample because of the tendency of lenses to magnify their images. Thus, since low luminosity objects are more numerous than bright ones, the lenses sample will be drawn from a larger population than the unlensed sample. The ratio, B, of these populations can be very large (i.e. TOG found B=26). For a finite core radius, this ratio is rather difficult to compute analytically since the probability of magnification P(M) and the minimum magnification M_{min} both depend on r_c. Namely, in this case the range over which l can vary and still result in multiple image formation depends on β and hence on lens size. Thus, this bias can change the shape of the distribution as well as its overall normalization. In general the effect will be to slightly ease the suppression in P (Δθ) for small Δθ.

The Results: (b) Dark Matter. Our results suggest that "dark" galaxies cannot play a significant role either in multiple image formation, or in affecting the image splittings of quasars lensed by galaxies. This is because it is generally believed (for example, see Blumenthal et al, 1986), that the dissipative collapse of baryons in galaxy formation tightens up the scale size of any associated dark halo. For halos without galaxies, which are presumably associated with systems with relatively small velocity dispersions (assuming visible galaxies result from rare, large primordial fluctuations), a large core radius would inevitably mean large β, and hence a small, or vanishing, cross section for multiple image formation.

Dark objects can, however, play a significant role in magnifying quasars without multiply imaging them. While this is impossible according to the standard wisdom, our analysis of the probability of amplification revealed that even though multiple image formation was cut off at values of l much smaller than a_{cr} for $\beta \neq 0$, the magnification for small β remained substantially the same all the way out to a_{cr}. Thus, the probability for magnification is not suppressed, while the probability for multiple image formation is! In this case, if an $\Omega=1$ population of dark objects exists, it could imply probabilities of order 1, in a flux limited sample (with large B) for image magnification. This could imply that such dark lenses alter the quasar luminosity function so that what we see does not correspond to the actual shape of this function. One way to probe for this effect might be to look for additional dependence of the quasar luminosity function on redshift in a way that could be analytically estimated in advance.

Conclusions and Future Work. Our results suggest that not only does the introduction of a finite core radius signficantly alter lensing statistics, but that a simple analytical model such as we have proposed can provide a basis for realistic predictions of the character of observations which will be made over the coming years. In this way, we may not only learn about the distribution and mass of galactic systems, but also about the possible existence of other, perhaps more abundant, lensing objects.

The work I have reviewed here suggests that further analytical studies of the effects of observational biasing, image magnification, and redshift dependence in this model could provide useful information on topics ranging from the need for dark matter in and outside of galactic systems to the intrinsic quasar luminosity function. In addition, use of this model might provide more realistic estimates of the statistics of new lensing observables such as Einstein rings, which have recently been reported. We hope, in the near future, to take up this call.

Acknowledgments: I have learned almost all I know about gravitational lensing first and foremost from Gary Hinshaw, who is responsible for most of what I have reported on here, and next from patient tutorials from B. Burke, J. Hewitt, I. Shapiro, and E. Turner.

References:

Anderson, N., and Alcock, C. 1986, Ap. J., 300, 56.
Avni, Y. 1981, Ap. J. (Letters), 248, L95.
Blumenthal, G.R., Faber S.M., Flores, R., and Primack, J.R. 1986, Ap. J. 301, 27.
Burke, B. F. 1984, Comments Ap., 10, 75.
Burstein, D., and Rubin, V.C. 1985, Ap. J., 297, 423.
Dyer, C. 1984, Ap. J, 287, 26.
Hewitt, J. 1986, PhD thesis, MIT
Hewitt, J. et al., 1988, these proceedings
Hinshaw, G., 1988, PhD thesis, Harvard University
Hinshaw, G. and Krauss, L.M., 1987, Ap. J. 320, 468
Gott, J. R.1985,Ap. J, 288, 422.
Ostriker, J. P., and Vietri, M. 1986, Ap. J. , 300, 68.
Peebles, P.J.E. 1986, Nature, 321, 27.
Schechter, P.L. 1976, Ap. J., 203, 297.
Turner, E. L. 1988, these proceedings.
Turner, E. L. 1980, Ap. J. (Letters), 242, L135.
Turner, E. L., Ostriker, J. P., and Gott, J. R. 1984, Ap. J., 284, 1 (TOG).
Tyson, J. A. 1981, Ap. J. (Letters), 248, 189.
Vietri, M. 1985, Ap. J., 293, 343.
Vietri, M. and Ostriker, J. P. 1983, Ap. J. , 267, 488.
Vilenkin, A. 1984, Ap. J. (Letters), 282, L51.
Young, P., Gunn, J. E., Kristian, J., Oke, J. B., and Westphal, J. A. 1980, Ap. J., 241,507.
Zwicky, F. 1937, Phys. Rev., 51, 290.

HIGHLY COLINEAR RADIO SOURCES AND CONSTRAINTS ON GRAVITATIONAL LENS SPACE DENSITY

Colin J. Lonsdale
Haystack Observatory
Off Route 40
Westford, MA 01886

Introduction. The point of this contribution is that one does not need to demonstrate positive evidence of gravitational lensing in order to say something useful about the distribution of clumped matter in the universe. Clearly, if we knew what the intrinsic (unlensed) structure of a high redshift object was, we could observe departures from that intrinsic structure, attribute the distortion to gravitational lensing, and deduce some, if not all, of the relevant parameters of the distribution of lensing matter. The standard approach, discussed at length in several papers in this volume, is to find a source that has clearly been strongly distorted (to the point of multiple imaging) and then find a reasonable model for both the intrinsic source structure and distribution of lensing mass that fits the data. A fine example of this approach is the demonstration by Roger Blandford (this volume) that the "Einstein ring" source (Hewitt *et al.* 1988) is well modelled by a fairly typical core–jet source lensed by an elliptical gravitational potential (such as might be found in an elliptical galaxy). What makes such analyses difficult is that we have little or no a priori knowledge of the intrinsic source structure, especially since radio sources at high redshift seem to be generally highly distorted (Barthel *et al.* 1988; Barthel and Miley 1988).

An alternative approach, described in this paper, is to find sources whose intrinsic structure we think we *do* know and set upper limits to the degree of lensing distortion we see. Highly colinear structure in a radio source is obviously not a guarantee that such colinearity is intrinsic to the source, since lensing can "straighten out" bent sources, but the key is that the lensing distortions should be random. It is reasonable, even theoretically expected, that some particularly undisturbed radio sources should be extremely straight, so if we find enough sources at high redshift that are straight enough, we can make a statistical argument that they are lensed by no more than some very small amount, thus constraining the matter distribution along the line of sight. We have found 2 sources out of a sample of perhaps 30 that are extraordinarily well aligned, and we claim that this is sufficient to place interesting limits on the density of clumped matter in the universe.

The Data. I have been mapping high redshift radio sources for several years using MERLIN and the VLA for reasons totally unrelated to gravitational lens studies. There is certainly no bias, in the sources that I have chosen, towards colinearity. Indeed, in the early days of the MERLIN instrument, with limited observing time and resources, there was a distinct tendency to look at the distorted, peculiar objects since they were more likely to be "interesting" astrophysically. I mapped perhaps 20 sources with MERLIN, another 85 as part of a major project on high redshift quasars with the VLA (preliminary 6-cm maps can be found in Barthel *et al.* 1988), and another 15 with the VLA in a study of multiple hotspots. Of these sources, most are so small that there are few resolution elements on the map, and therefore there exists no possibility of accurately determining the alignment of the sources. Many more sources lack compact hotspots in both outer lobes, also precluding alignment measurements. The remainder have compact features in both lobes and more than 30 resolution elements across the maps. Thus, I have a heterogeneous sample of about 30 sources whose degree of colinearity can be accurately determined and in which I would have noticed extreme straightness.

Of these 30 sources, two (0835+580 = 3C205, and 0922+149) are spectacularly well aligned. In both sources, there is more than one peak of emission in one of the lobes, but in both cases there is only one truly unresolved hotspot. A variety of properties clearly identifies the compact hotspot as the primary feature in the lobe (see Lonsdale and Barthel 1986), and the alignment numbers quoted below refer to these compact hotspots. Using the definition of the bending angle ϕ illustrated below, for 3C205, $\phi \leq 0\overset{\circ}{.}15$, and for 0922+149, $\phi \sim 0\overset{\circ}{.}1$. The positions of the hotspots and cores were determined using the IMFIT model-fitting program in the NRAO AIPS software package, and the instrumental/mapping/fitting uncertainties are likely to introduce errors at the level of $\sim 0\overset{\circ}{.}1$ in ϕ.

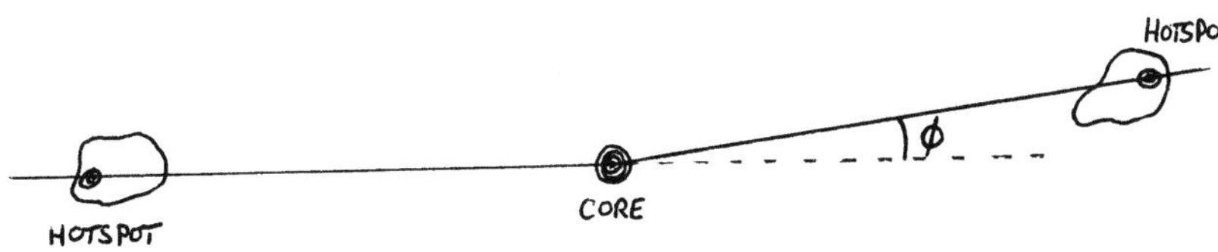

Discussion. Blandford and Jaroszyński (1981) examined the distortion of triple radio sources by the gravitational lensing from matter clumped on scales of galaxies to clusters of galaxies. They derived the following formula for the rms bending angle one should expect under reasonable assumptions:

$$\langle\phi^2\rangle^{1/2} = 160^\circ(\theta/1'')^{-0.4}\overline{\Omega}S^{1/2}(z,\Omega) \ ,$$

where

$\phi =$ bending angle in degrees,

$\theta =$ source size,

$\overline{\Omega} =$ density of matter in large-scale clumps (in units of the closure density),

$\Omega =$ total density of matter,

$S =$ function dependent on Ω and correlation function of clumps.

This formula is valid for $10'' \leq \theta \leq 60''$ and $z >\sim 1$. For 3C205, $\theta = 17''$ and $z = 1.53$, yielding a value for $\langle\phi^2\rangle^{1/2}$ of around $13^\circ\overline{\Omega}$. The corresponding value for 0922+149 ($\theta = 40''$, $z = 0.9$) is around $9^\circ\overline{\Omega}$.

Thus, the expected rms bending angle $\langle\phi^2\rangle^{1/2} \sim 100\overline{\Omega}\phi_{\text{observed}}$ for *both* sources. If the distribution of ϕ has a half-width of several degrees, due either to lensing or intrinsic bending, these two sources represent a statistically significant spike at the origin, which must be explained. For example, if the bending angles are randomly distributed between 0° and $\sim 15^\circ$ ($\langle\phi^2\rangle^{1/2} \sim 9^\circ$), the probability of two sources lying so close to 0° is about 0.02. This distribution is comparable to both the observed distribution of bending angles in the 30 sources, as well as the expected bending angle distribution due to lenses if $\Omega = 1$.

The proper way to interpret these data is to regard the observed distribution of bending angles in high redshift radio sources as the convolution of the intrinsic bending angle distribution and the bending angle distribution due to gravitational lensing, as characterized by Blandford and Jaroszyński (1981). If *either* distribution is smooth and broad, it is impossible to obtain the observed spike. We know little about the intrinsic bending angle distribution, but it would not be surprising if it contained a spike at the origin, caused by a few straight, undisturbed sources whose ejection axis is stable (indeed, the present data require that such a spike be present). A signature of intrinsic bending is a linear polarization E-vector angle that remains perpendicular to the ridge line of a bend, since this position angle is unaffected by lensing. Such "bend tracking" behavior is common among these sources, implying that a major portion of the breadth of the distribution is intrinsic in origin.

The data also require that the gravitational lens component of the bending angle distribution be either narrow or spiky. Spikiness is precluded by the assumptions made by Blandford and Jaroszyński (1981) regarding the range of size scales for the clumping. If the universe is uniform on scales larger than their largest clumping scales, the optical depth to gravitational deflections becomes large at $z \sim 1$, and the analysis is valid, producing a smooth distribution with the above rms value. However, if the universe is clumped on still larger scales, the probability that the line of sight to any given source misses *all* the lenses becomes significant (the optical depth drops to a few or less), and a spike develops in the lensing bending angle distribution at the origin.

Conclusions. There are two conclusions to be drawn from this study. First, the intrinsic bending angle distribution of radio sources at high redshifts has a spike at the origin (i.e., some radio sources are just intrinsically *very* straight). Second, and of relevance to this conference, the lens-induced bending angle distribution *also* has a spike at the origin. This may be due to two causes. The distribution may simply be narrow, which by reference to the above calculations implies that $\overline{\Omega}$ must be fairly small. A value of perhaps 0.3 would reduce the significance of these two sources to a reasonable level, ignoring the likelihood of a broad (if spiky) intrinsic distribution. Alternatively, the universe could be clumpy on truly vast scales, so that we are looking at these two sources along lines of sight that are simply deficient in clumped matter. Note that these conclusions get stronger if many sources are intrinsically bent.

References.

Barthel, P.D., and Miley, G.K. 1988, *Nature*, **333**, 319.

Barthel, P.D., Miley, G.K., Schilizzi, R.T., and Lonsdale, C.J. 1988, *Astr. Ap. Suppl.*, **73**, 515.

Blandford, R.D., and Jaroszyński, M. 1981, *Ap. J.*, **246**, 1.

Hewitt, J.N., Turner, E.L., Schneider, D.P., Burke, B.F., Langston, G.I., and Lawrence, C.R. 1988, *Nature*, **333**, 537.

Lonsdale, C.J., and Barthel, P.D. 1986, *A. J.*, **92**, 12.

VII. Microlensing

GRAVITATIONAL MICROLENSING

William D. Watson
Departments of Physics & Astronomy
University of Illinois at Urbana-Champaign
1110 West Green Street, Urbana, IL 61801 USA

Abstract. Gravitational lensing of distant light sources in the universe by compact masses M that are approximately stellar -- $10^6\ M_\odot \gtrsim M \gtrsim 10^{-4}\ M_\odot$ -- is reviewed. In the lensing by galaxies and especially at optical and X-ray wavelengths, the effect of individual stars or the "granularity" in the matter distribution for galaxies can reasonably be expected to be significant. Though microlensing may well underlie various observations, there seem as yet to be no generally accepted cases in which it has been unambiguously identified. Recent interpretations of the association between galaxies and X-ray selected AGNs are promising, however.

Introduction. "Gravitational microlensing" ordinarily refers to the lensing effects of individual stars or other compact masses, typically with masses $10^6\ M_\odot \gtrsim M \gtrsim 10^{-4}\ M_\odot$, which may be superimposed on the lensing by an entire galaxy. That is, microlensing can be viewed as the effect of the deviation from a uniform distribution of matter (or the "granularity") that occurs because the mass of galaxies is (at least partly) in the form of compact masses. Chang and Refsdal (1979) seem to have first recognized that the relevant astronomical parameters are such that microlensing may reasonably be expected to play a role in the gravitational lensing by galaxies.

The characteristic length scale for microlensing is the Einstein radius a_0 for the single mass. For significant gravitational lensing by an individual compact mass, the rays of light must pass near the mass -- within a distance roughly equal to its Einstein radius. For cosmological distances, the Einstein radius is roughly

$$a_0 \simeq 5 \times 10^{16} \sqrt{M/M_\odot}\ \text{cm} \tag{1}$$

which corresponds to an angle

$$\theta_o \simeq 10^{-6}\sqrt{M/M_\odot}\ \text{arc sec} \qquad (2)$$

for a compact lensing mass M in terms of the solar mass $M_\odot$.

If a large fraction of the light from a distant source is to be affected by lensing due to the single mass, a similarly large fraction of the rays must pass within a_o of the mass. Hence, the requirement for the angular size of the source θ_s is

$$\theta_s \lesssim \theta_o \qquad (3)$$

for significant microlensing. The central, optical continuum emitting regions and the X-ray emitting regions of quasars may satisfy eqn. (3) for $M \simeq M_\odot$. In contrast, there is no evidence that the emitting regions of extragalactic radio sources can be small enough to satisfy eqn. (3) unless $M > 100\ M_\odot$ [if lensing masses within our own galaxy are considered, θ_o is increased to approximately a milli-arc second but the probability for lensing along a randomly chosen line of sight is small ($\simeq 10^{-5} - 10^{-6}$)].

An attractive aspect of microlensing is the prediction of time variations. That is, its magnification effect should change on a time scale of roughly

$$t_m = a_o/v_\perp \qquad (4)$$

where $v_\perp$ is the characteristic relative velocity perpendicular to the line-of-sight between the source, observer and the lensing mass. Due to galactic rotation $v_\perp$ should be at least 300 km/s and may reasonably be about 1000 km/s in many cases due to the random motion of galaxies. Time scales $t_m \simeq 10$ yrs are then short enough to be sought even for $M \simeq M_\odot$. The characteristic "rise time" for the magnification is roughly

$$t_r \simeq a_s/v_\perp \qquad (5)$$

and can be much shorter than t_m when $a_s << a_o$. It may also be a useful signature of microlensing.

On the other hand, since θ_o is the characteristic angular size for any changes of the image due to microlensing, multiple images due to microlensing cannot be expected to be resolved. From an observational viewpoint, the effect of microlensing is thus simply to alter the apparent luminosity of the image.

Although the effect of microlensing in enhancing the apparent luminosity of a source can be comparable to that of the entire galaxy, the unambiguous identification of microlensing can be expected to be difficult and this has so far proved to be the case. Impressive evidence for microlensing has recently been reported (see Section II) and it may well withstand future scrutiny. Because the relevant time scales tend to be measured in years and gravitational lenses have been detected for only a few multiples of this time scale, it can reasonably be expected that data from the monitoring of lensed images will begin to yield a significantly improved assessment of the astronomical impact of microlensing. Numerous astronomical applications for microlensing have been recognized; some of these will be summarized in Section II. In Section III, somewhat more detailed results will be discussed about the calculation of the effects of microlensing when a number of stars contribute and the single mass approximation is not valid -- a situation that is expected to be common.

II. Astronomical applications of microlensing. In their original proposal, Chang and Refsdal (1979) pointed out that there may be uncorrelated variations in the apparent luminosity of the A and B images of the gravitational lens system QSO 0957 + 561 due to the influence of microlensing. This might lead to confusion in attempts to determine the Hubble parameter by measuring the time delay between the luminosity variations in the two images that result from a single brightness variation of the source (Refsdal 1964). Microlensing depends only on the stars near the path of light, and the paths of the light for separate images due to lensing by the entire galaxy pass through widely separated locations within the galaxy. The reported, uncorrelated light variation in the lensed images of QSO 1115 + 080 (Foy, Bonneau and Blazit 1985) may be a manifestation of this aspect of microlensing. Not only can microlensing cause confusion in recognizing the correlated variations in the two images, but it also alters the flux from that which would be found for a smooth distribution of matter. Inferring the mass of the galaxy and other parameters of the lensing system (also entering into a determination of the Hubble parameter utilizing gravitational lenses) which utilizes the apparent luminosities of the two images (Refsdal 1964), then contains a statistical uncertainty due to the granularity of the matter of the galaxy. It follows immediately (Gott 1981; Young 1981; Canizares 1982) that the predicted light variations due to microlensing can, in principle, be utilized to search for dark matter that may be in the form of nonluminous compact objects such as "Jupiters," brown dwarfs, black holes, etc.

The rapid rise time t_r has been proposed by Grieger, Kayser and Refsdal (1988) as a probe for the internal structure of the quasars. For two observers at locations that are separated by distances that are characteristic of the solar system, the increase in flux due to microlensing will occur at slightly different times. The variations in flux that occur as a source with structure passes across the caustics will then occur at different, but correlated, times for the two observer. Such variations could then be distinguished from intrinsic variations of the source.

Variations in the apparent luminosity of active galactic nuclei might be caused by microlensing. In particular, the observed enhancement of the central continuum in comparison with the larger, spectral line emitting region might arise because the central, continuum region in the unlensed source is comparable or smaller than the Einstein radius and is thus magnified (Ostriker and Vietri 1985) whereas the line emitting region is larger than the Einstein radius.

One of the earliest (Canizares 1981; Vietri and Ostriker 1983; Peacock 1983) as well as currently the most exciting astronomical applications of microlensing is its potential to alter the observed distribution of quasars -- in both space and apparent brightness. The possibility that there is a slight statistical excess of quasars projected on the sky near galaxies has been discussed for some time. What is new is that based on utilizing X-ray selected quasars, Stocke *et al.* (1987) have reported a statistically impressive association between these objects and bright galaxies that is reproduced in Figure 1. This has been interpreted as suggesting that microlensing may reasonably be responsible for a significant alteration in the distribution of the apparent luminosities for quasars. Previous objections to such an alteration have, in part, been based on the premise that a lensing-dominated distribution must yield $d[\log N(S)]/d[\log S] = -2$ for the number of quasars N(S) at redshift z with flux > S. These objections now seem less serious as a result of the demonstration that the effect of finite source size is to allow other predicted slopes (Schneider 1987; see Figure 2).

<u>III. Calculation of microlensing.</u> Treating microlensing, or the effect of granularity in the matter distribution of galaxies, in terms of lensing by isolated compact masses seems to be adequate only when the surface density of matter of the lensing galaxy is quite small. When the "optical depth" τ for gravitational lensing ($\tau = \pi n a_0^2$ where n is the surface number density of compact masses) is greater than a few tenths, the simultaneous lensing by a number of masses must be calculated. Questions to be answered by such calculations are, how rapid are the time variations in the observed flux? How do the results depend upon the size of the source, upon the amount of matter in compact masses and upon the amount of matter in a completely smooth distribution (gas, neutrinos, etc)?

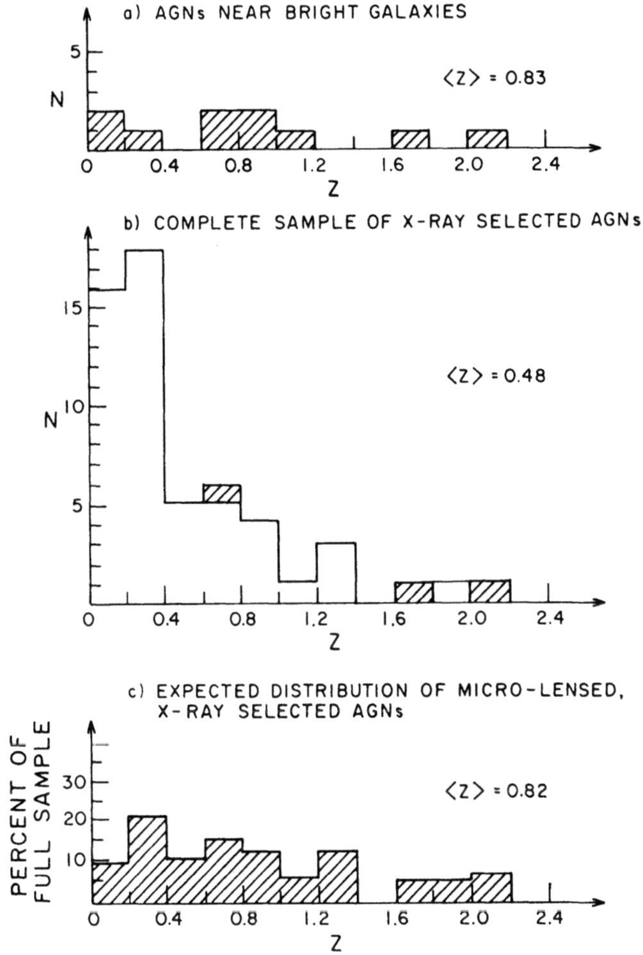

Figure 1. The redshift distributions, for a statistical sample of X-ray-selected AGNs near bright galaxies in Fig. 1a and for a complete sample of X-ray-selected AGNs in Fig. 1b. The distribution of redshifts shown in Fig. 1c is that expected due to microlensing. The lack of a statistically significant difference between (a) and (c) is taken to indicate that the unusual redshift distribution for X-ray-selected AGNs near galaxies can be explained by the effects of microlensing (from Stocke *et. al* 1987).

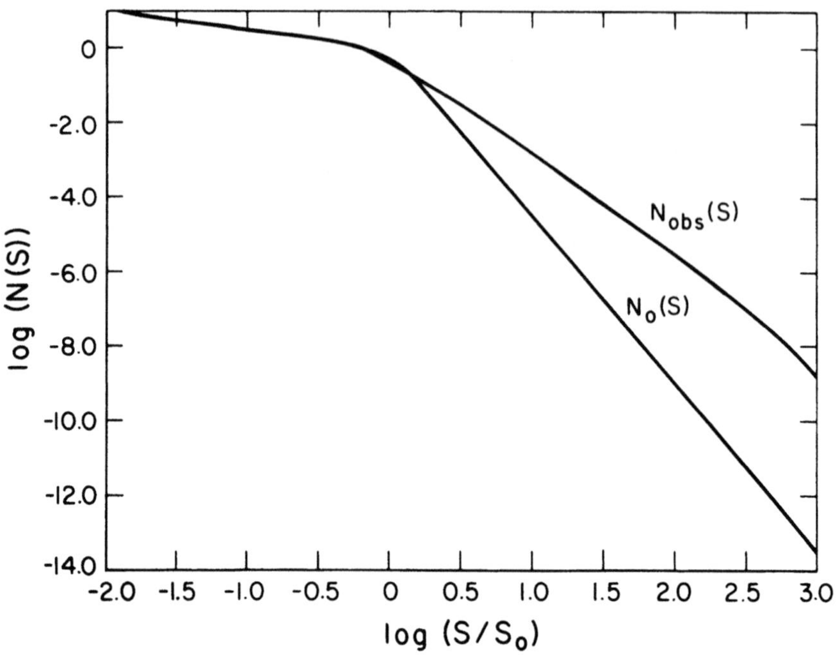

Figure 2. Source counts N(S) for fluxes greater than S for redshift z = 2 versus normalized flux. $N_o(S)$ is the distribution in the absence of lensing and $N_{obs}(S)$ is the distribution that results from microlensing for a particular choice of finite source size and range of lensing masses. This demonstrates that lensing can lead to a slope different from -2 (from Schneider 1987).

The initial, limited numerical computations using Monte Carlo type methods due to Young (1981) have been criticized for limited sampling, and several investigators have performed further computations (also of the Monte Carlo type). Paczynski (1986) examined the case of the microlensing of a point source and obtained the time variation in the flux due to the random motion of the compact masses. At any time, there are numerous images of the source that are separated by angles that typically are measured in micro-arc seconds (for 1 $M_\odot$ masses) as discussed in Section I. A representative case is reproduced in Figure 3. The finite size of the source is expected to be an important consideration. The analogous, predicted variations in flux for the microlensing of finite-size sources have been obtained for $\tau < 0.4$ by Kayser, Refsdal and Stabell (1986) and extended to larger values of τ by Schneider and Weiss (1987).

Representative light curves are reproduced in Figure 4. The sensitivity to source size, the rapid rise time for small sources, and the longer characteristic length scale a_0 for variations are evident. Another feature is the dependence of such light curves on the surface mass density as measured by τ. From the computations of Schneider and Weiss (1987), the variations (in astronomical magnitudes) are larger and qualitatively similar for τ both larger and smaller than one. As τ approaches one, the variations become smaller in amplitude and more rapid.

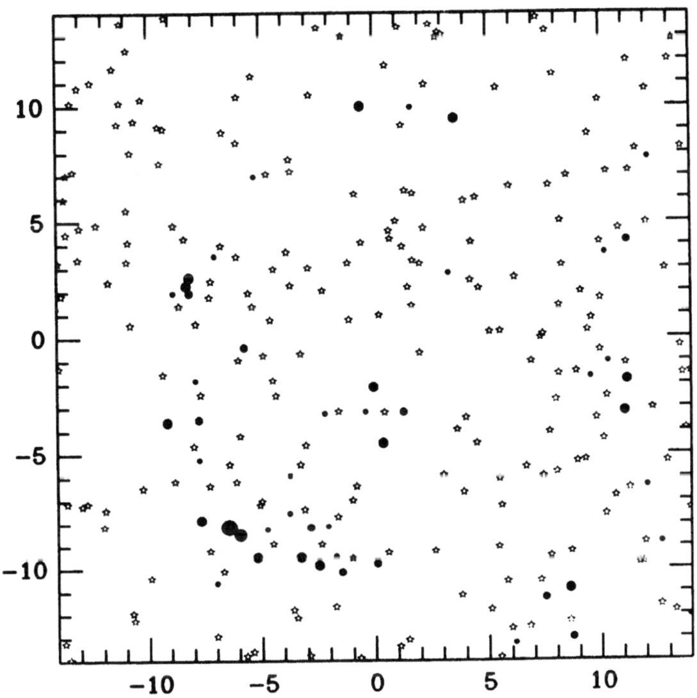

Figure 3. Representative locations (in angle) of images due to the microlensing of a point source by randomly distributed stars (indicated by starlike marks). The scale is micro-arc seconds for a cosmological distance. Locations of the brightest microimages are given by filled circles; the faintest microimages (smallest circles) are approximately 100 times fainter than the brightest microimages. The point source is at the center and there are about as many images as stars (from Paczynski 1986).

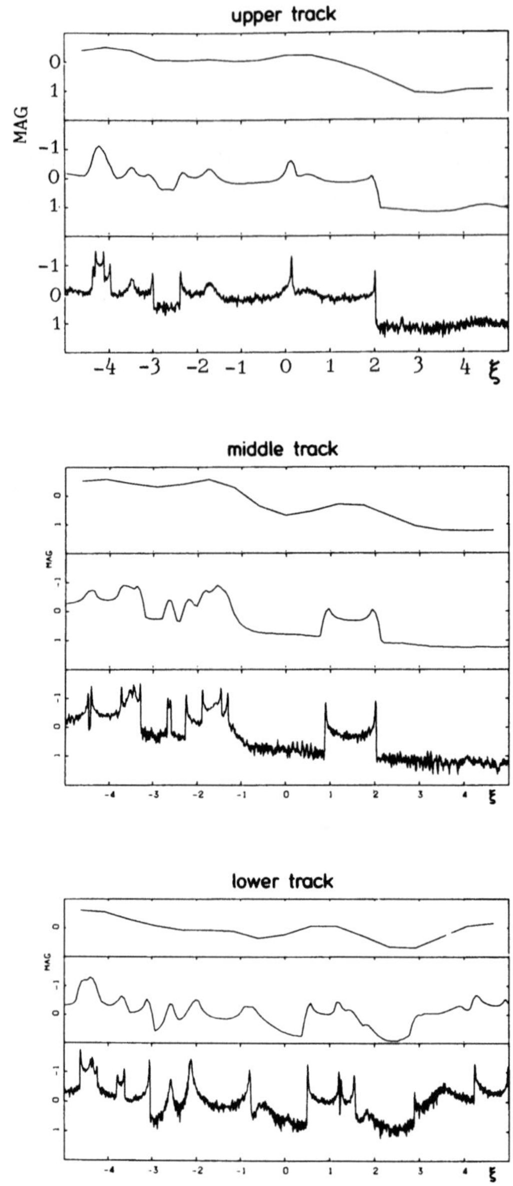

Figure 4. Representative changes in the total flux from all microimages (magnitudes) versus distance moved by an observer perpendicular to the line-of-sight (units are roughly a_0 for typical parameters). "Track" refers to the choice of paths across the distribution of lensing masses. The three curves for each track are for source sizes of approximately 1, 0.1 and 0.01 a_0. Optical depth τ = 0.4 (from Kayser *et al.* 1986).

Whereas the foregoing calculations are of the numerical, Monte Carlo type, S. Deguchi and I (Deguchi and Watson 1987b) developed a largely analytic, statistical treatment of microlensing by a collection of masses valid for all τ and for sources of finite size. The key aspect of this calculation is in utilizing a "Markoff" method to perform the multiple integrals over the possible positions of the compact masses. From a calculation of this type, statistical properties (such as mean square luminosities, power spectra and autocorrelation functions) are obtained, not actual (predicted) light curves as in the simulations of the Monte Carlo type. In Figure 5, we reproduce the mean square luminosities obtained in this treatment. The magnitudes and general properties displayed in Figure 5 do seem to be similar to those found in the numerical simulations.

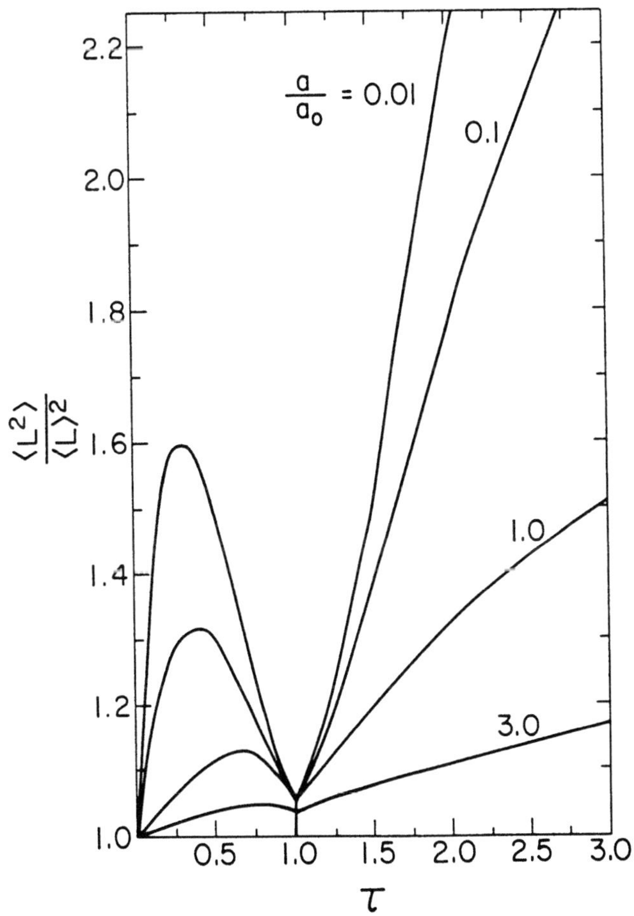

Figure 5. Mean square of the total flux L, which measures the variation in flux due to microlensing versus optical depth τ for lensing. The ration a/a_0 is source size/a_0 (from Deguchi and Watson 1987b).

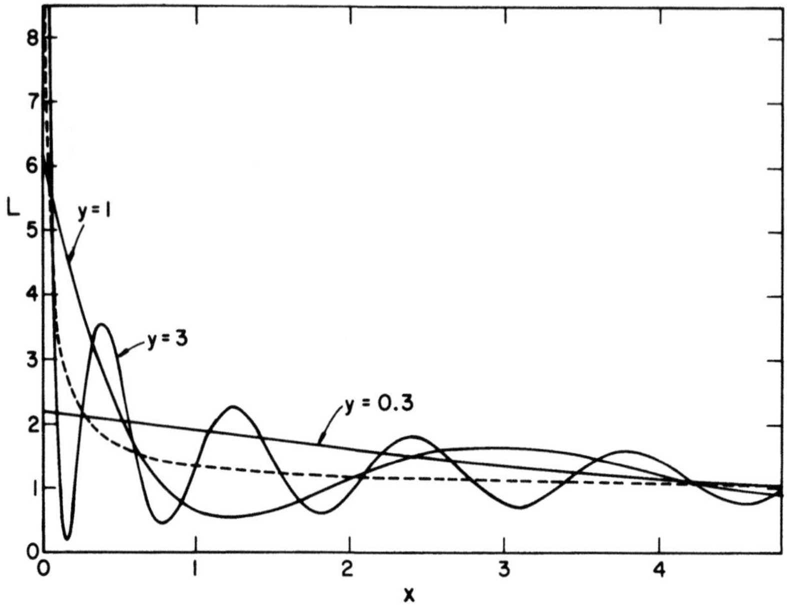

Figure 6. The enhancement L of the flux from a point source due to gravitational lensing by a single point mass. The distance X (dimensionless units) is the location of the observer measured perpendicular to the line joining the source and the lensing mass. The parameter Y = 2π times (Schwarzschild radius/wavelength of light) = 2×10^6 $(M/M_\odot)/\lambda$(cm) (from Deguchi and Watson 1986a).

The approximation of geometrical optics underlies the basic "ray equations" used to treat gravitational lensing. When the smaller masses of individual objects (as opposed to galaxies) are considered in microlensing, with the resultant smaller splitting for images and hence smaller differences in paths for the various contributing rays, the wave effects (i.e., diffraction) must enter at some level. The ability to detect wave effects seems to have first been considered by Mandzhos (1981). S. Deguchi and I (Deguchi and Watson 1986a,b) have considered it further and in considerable detail. For a point source, the effect can be large as shown in Figure 6. It is much smaller and would be quite difficult to detect for sources of finite size, especially when the possible influence of free electrons is considered in the medium through which the waves are propagating (Deguchi and Watson 1987a).

The author's research is supported in part by the National Science Foundation.

References

Canizares, C. R. 1981, *Nature*, **291**, 620 (errata **293**, 490).

Canizares, C. R. 1981, *Ap. J.*, **263**, 508.

Chang, K., and Refsdal, S. 1979, *Nature*, **282**, 561.

Deguchi, S., and Watson, W. D. 1986a, *Ap. J.*, **307**, 30.

Deguchi, S., and Watson, W. D. 1986b, *Phys. Rev.* D, **34**, 1708.

Deguchi, S., and Watson, W. D. 1987a, *Ap. J.*, **315**, 440.

Deguchi, S., and Watson, W. D. 1987b, *Phys. Rev. Lett.*, **59**, 2814.

Foy, R., Bonneau, D., and Blazit, A. 1985, *Astron. Ap.*, **149**, L13.

Gott, J. R. 1981, *Ap. J.*, **243**, 140.

Grieger, B., Kayser, R., and Refsdal, S. 1988, *Astron. Ap.*, **194**, 54.

Kayser, R., Refsdal, S., and Stabell, R. 1986, *Astron. Ap.*, **166**, 36.

Mandzhos, A. V. 1981, *Sov. Astr. Lett.*, **7**, 213.

Ostriker, J. P., and Vietri, M. 1985, *Nature*, **318**, 446.

Paczynski, B. 1986, *Ap. J.*, **301** 503.

Peacock, J. 1983, in *Quasars and Gravitational Lenses: Proc. 24th Liege Int. Ap. Colloq.* (Liege: Universite de Liege), p. 98.

Refsdal, S. 1964, *M.N.R.A.S.*, **128**, 307.

Schneider, P. 1987, *Ap. J. Lett.*, **316**, L7.

Schneider, P., and Weiss, A. 1987, *Astron. Ap.*, **171**, 49.

Stocke, J. T., Schneider, P., Morris, S., Gioia, I., Maccaro, T., and Schild, R. 1987, *Ap. J. Lett.*, **315**, L11.

Vietri, M., and Ostriker, J. P. 1983, *Ap. J.*, **267**, 488.

Young, P. 1981, *Ap. J.*, **244**, 756.

COSMIC DENSITY ESTIMATE FROM MICROLENSING

H. W. Rix and C. J. Hogan
Steward Observatory
University of Arizona
Tucson, AZ 85721 USA

Introduction. Analyzing a flux-limited, statistically complete sample of 200 X-ray selected AGN's, Stocke *et al.* (1987) found that the subsample consisting of the 10 AGN's with the nearest projected foreground galaxies ($V_{\rm gal} < 18$ mag) has a substantially higher mean redshift than the rest of the sample (97.5% confidence). They hypothesized, based on the distinguishing criterion for the subsample, that the apparent overluminosity of the subsample members was caused by microlensing amplification in the environment of the foreground galaxies.

Motivated by this claim, we developed a method to use this type of data to make an estimate of the cosmic density parameter Ω (Rix and Hogan 1988). Our derivation is based on very simple assumptions and has the advantage that the resulting estimate of Ω is relatively insensitive to the critical assumptions (e.g., the slope of the AGN luminosity function). In this present case, the microlensing hypothesis yields a formal lower limit on the mean density of pointlike lensing objects of $\Omega = 0.25$ at 95% confidence.

Estimate of Ω. As a first step, we estimate how much amplification is implied statistically by the Stocke *et al.* (1987) sample. The observed flux from each AGN is related to its intrinsic flux via

$$F_{\rm observed} = L_{\rm intrinsic} \times \mu_a \times d_l^{-2}$$

where μ_a is the flux amplification due to hypothetical microlensing and d_l is the luminosity distance. Averaging these quantities over both the sample with foreground galaxies (index "G") and the control sample ("C"), one obtains for the mean intensity enhancement $\epsilon \equiv \mu - 1$ of the G-sample:

$$\frac{\langle\mu\rangle_G}{\langle\mu\rangle_C} - 1 = \langle\epsilon\rangle_G = 4.4 \pm 2.2$$

However, the high statistical significance of this result (96% confidence level) vanishes if any two of the five apparently brightest objects are removed from the G-sample. Therefore, in order to explain the statistical peculiarities of this sample by microlensing, we need a density Ω sufficient to raise the probability that microlensing causes substantial enhancement ($\epsilon \approx 1$–5) in about 2 out of 200 cases above some value, determined by the required likelihood or confidence level.

To assess this probability we make the following assumptions: (1) the AGN's are infinitely distant point sources (i.e., $z_{\rm AGN} \gg z_{\rm galaxy}$), (2) the lensing is by a point mass at the redshift of the foreground galaxy, and (3) one single lensing event dominates the amplification.

No model of galactic halos needs to be made. However, a flux-limited sample requires the consideration of amplification bias because amplified sources have a better chance of entering the sample. With these assumptions, the probability Φ for a particular incidence of lensing at distance z_i resulting in an enhancement of at least ϵ_i is

$$\Phi = \frac{3}{2} \cdot z_i{}^2 \cdot P_b(\epsilon_i) \cdot \Omega \quad ,$$

where $P_b(\epsilon)$ is the normalized integrated probability for biased amplification of more than ϵ in a flux-limited sample. For the observed AGN X-ray luminosity function, $P_b(\epsilon)$ can be computed and, for $\epsilon \geq 1$, it does not depend sensitively either on ϵ_i or the slope of the luminosity function. Multiplying these Φ's for a complete flux-limited sample gives the joint probability for n amplification events (out of N) as

$$L = \left(\frac{3}{2}\Omega_l\right)^n \prod_{i=1,n} [z_i{}^2 P_b(\epsilon_i)(N+1-i)] \quad .$$

Since we cannot determine ϵ_i in individual cases but at best point out a few good candidates for high amplification ($\epsilon > 1$), we cannot calculate the complete likelihood distribution. However, it is possible to place formal upper limits on Ω_l from the likelihood of the apparent overluminosity being due to lensing, as shown in the Table 1. This is done by identifying n cases with $\epsilon > 1$ (e.g., from Table 2), using the observed foreground galaxy redshifts for z_i, and evaluating that Ω_l for which this much amplification would be expected in a fraction L of realizations.

Table 1
Likelihood Limits on Ω_l for Two Events with $\epsilon = 5$

Real cases	Ω_l ($L = 1\%$)	Ω_l ($L = 5\%$)	Ω_l ($L = 10\%$)
1+2	0.18	0.40	0.57
1+7	0.85	1.9	2.7
1+8	0.73	1.6	2.3
2+7	0.13	0.29	0.42
2+8	0.11	0.25	0.35
7+8	0.54	1.2	1.7

Table 2
Properties of Microlens Candidates and Control Samples

No.	Name	z_q	z_g	$(Fd_l^2)_x$	$(Fd_l^2)_{opt}$	R
1	0038.8–0159	1.69	0.017	3.97	190	1.4
2	0104.2+3153	2.03	0.111	6.98	46	0.9
7[a]	0950.2+0804	1.45	0.023	2.27	—	5.0
8	1109.3+3544	0.91	0.027	1.06	12	0.7
mean of G-sample		0.83	0.039	1.43	33	1.8
median of G-sample		0.63	0.02	0.4	12	1.7
mean of control sample		0.48	–	0.28	9	–

[a] 7 is not in the G-sample because $R > 3$

Discussion. We investigated the statistical peculiarities of a statistically complete, flux-limited, X-ray selected sample of AGN's (Stocke *et al.* 1987). The probability that the apparent brightening of the AGN's with foreground galaxies is due to statistical noise is small (a few percent). Since microlensing is a plausible explanation for the apparent statistical brightening, we have investigated the implications of the lensing hypothesis for the cosmic density of lensing objects. We were able to relate the statistical characteristics of the sample to Ω_l in a rather straightforward way, obtaining very high values for Ω_l, namely $\Omega_l > 0.25$ on a 95% confidence level. However, the interpretation of this likelihood becomes problematic if additional a priori constraints on Ω_l (i.e., upper limits from other observations) are imposed. Therefore, that a statistical "fluke" is responsible for the apparent differences between the subsamples cannot be ruled out. We are currently enlarging the G-sample by a CCD imaging survey in order to improve the estimate of Ω and reduce statistical uncertainty.

Acknowledgments. This work was supported in parts by NASA grant NAGW-763 (H.W.R.) and NSF grant AST-8714667 (C.J.H.). We also acknowledge support from a Fulbright Scholarship (H.W.R.) and the Alfred P. Sloan Foundation (C.J.H.).

References.

Rix, H. W., and Hogan, C. J. 1988, *Ap. J.*, **332**, 108.

Stocke, J. T., Schneider, P., Morris, S. L., Gioia, I. M., Maccacaro, T., and Schild, R. E. 1987, *Ap. J. (Letters)*, **315**, L11.

MICRO-LENSING MODEL FOR QSO 2237+0305

J. Wambsganss, B. Paczyński and N. Katz
Princeton University Observatory
Peyton Hall, Princeton, NJ 08544 USA

Abstract. We have calculated micro-lensing effects for the four images of QSO 2237+0305 using the macro-lensing model developed by Schneider *et al.*(1988; STGHSL), with proper surface density and shear for every image. Our computations were done with a new code that makes use of the hierarchical tree method. The high efficiency of this approach allowed us to include a very large number of lensing stars with different masses. Our results show that the observed changes in luminosity of the four images of QSO2237+0305 may be interpreted as micro-lens induced.

Introduction. Many different micro-lens calculations have been performed, e.g., Schneider and Weiss (1987; SW); Kayser, Refsdal, and Stabell (1986; KRS); Paczyński (1986); Young (1981). In most cases, these papers present parameter studies with different combinations of external shear γ and surface density σ (here always in units of the critical surface density), and they assume uniform masses for all lensing objects. The number of lensing objects is usually of the order of 100, but with an improved technique SW used up to 8000 lensing stars. The region covered in the source plane is usually of the order of, or less than, 10 ξ_0 (ξ_0 = Einstein radius in the source plane), possibly too small to cover the largest features of the light curve. Most authors used circular sources with constant surface brightness.

QSO 2237+0305 is a gravitational lens system with the lensing galaxy at a redshift $z_L = 0.0394$ and the source at $z_S = 1.695$. Four images of the source have been found (STGHSL; Yee, 1988), typical separations are of order $1''$. All four images are very close to the center of the lensing galaxy.

Our objective was to model the micro-lens behavior of the four images of QSO2237+0305. We adopted the macro-lens model of STGHSL and used their values of surface density and shear for the four images (see Table 1), allowing us no freedom in our choice of parameters. Due to a new code that makes use of the hierarchical tree method, we were able to make several improvements in our calculations compared to other authors.

Table 1: Parameters for the Four Images of QSO 2237+0305

Image	A	B	C	D
Surface density σ	0.35	0.44	0.86	0.60
Shear γ	0.41	0.28	0.47	0.67
Amplification $A = [(1-\sigma)^2 - \gamma^2]^{-1}$	3.9	4.3	-5.0	-3.5
Number of lensing masses	29956	30397	58470	51161
Number of light rays (in 10^7)	3.09	2.19	1.63	2.03

Method. For our calculations, we used the inverse ray tracing method that is described elsewhere (SW, KRS). The receiving area was a circle with a diameter of 40 ξ_0 (ξ_0 = Einstein radii in the source plane for a 1 $M_\odot$ star), which corresponds to a length of 1.67 pc (for $H_0 = 75$, $q_0 = 0.5$). As far as we know, all other authors have used smaller areas, but the large coherent structures in Figure 1a show that a receiving area of at least this size is necessary in order to cover typical patterns. Our resolution was (500 pixels)^2, where one pixel length corresponds to 3.35×10^{-3} pc in the source plane (for the same values of H_0 and q_0). The lens masses were chosen in the mass interval $0.1 < m/M_\odot < 1$ according to a Salpeter mass function $[f(m)dm = m^{-2.35}dm]$ in order to model a realistic mass distribution of an old stellar population in the center of a galaxy. We considered up to 60,000 stars acting as lenses (see Table 1), about an order of magnitude more than other authors (SW). These stars were randomly distributed in the lens plane inside a circle whose radius was determined to be much larger than the shooting and the receiving circles. The shooting circle was chosen so that everywhere inside the receiving area we got more than 99% ray efficiency (see Katz, Balbus, and Paczyński 1986 or SW). The calculation of the deflection angles of the individual rays were performed with a code that made use of the hierarchical tree method (Barnes and Hut 1986). This code will be described elsewhere. By using this tree type code our gain in computing efficiency was more than factor of 100 for this number of lenses, compared to a brute force calculation including every lensing star individually.

Results: Amplification Patterns, Light Curves, and Correlation Functions. Because of lack of space we restrict ourselves in this paper to one case only, namely, image B. In Figure 1a the distribution of the amplification factor for different positions in the source plane is plotted for image B ($\sigma = 0.44$, $\gamma = 0.28$). For comparison, this distribution, for the same surface density,

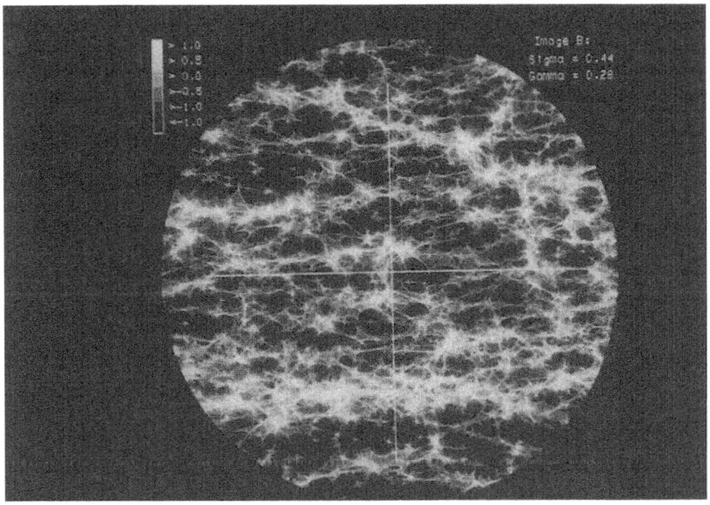

Figure 1a. Amplification pattern as function of the source position for image B ($\sigma = 0.44$, $\gamma = 0.28$). The steps in grey correspond to half a magnitude, respectively.

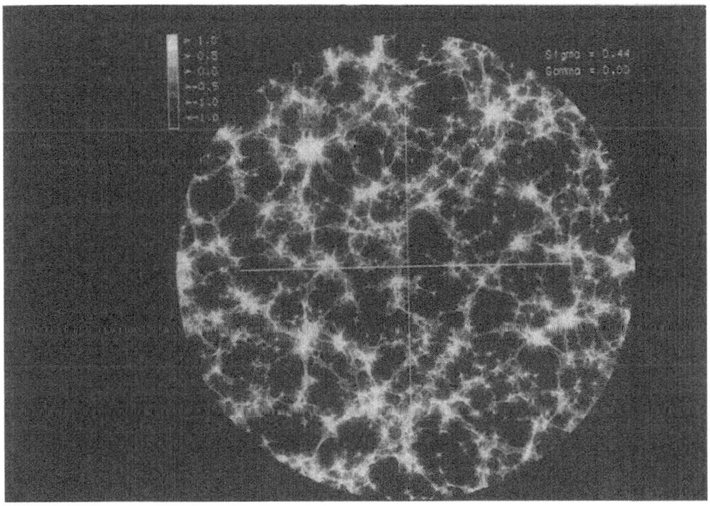

Figure 1b. Same as Fig. 1a without external shear ($\sigma = 0.44$, $\gamma = 0.0$).

but without external shear ($\gamma = 0$) is shown in Figure 1b. The differences in terms of isotropy are obvious.

For the light curves, we used Gaussian luminosity profiles $[L(r) \propto e^{-r^2/2\sigma_l^2}]$ with different half widths $\sigma_l = 1, 2, 4, 8, 16$ pixels, corresponding to source half widths of 3.35, . . . , 53.6×10^{-3} pc, respectively (for $H_0 = 75$, $q_0 = 0.5$). We assumed source and observer to be fixed and the intervening galaxy to move with a center-of-mass velocity of 600 km/sec. For simplicity, we did not consider random velocities of individual stars. For a detailed investigation of the different velocities, see KRS. We present two light curves L(t) for image B, one parallel to the action of the shear (Fig. 2a) and one in the orthogonal direction (Fig. 2b). The lines along which the light curves are calculated are indicated in Figure 1a.

The light curves show the typical behavior for micro-lensing but it is difficult to quantify the degree of agreement or disagreement between an observed and a calculated light curve. Therefore we calculated an auto-correlation function for the light curves, which can more easily be compared with observational data. The normalized auto-correlation function was computed according to the equation

$$f(\Delta t) = \frac{\langle L(t) \times L(t+\Delta t) - \langle L \rangle^2 \rangle}{\langle L^2 \rangle - \langle L \rangle^2} \quad .$$

For the five different half widths (1, 2, 4, 8, 16 pixels) of image B, the results are shown in Figures 3a,b. These curves reflect the strong/weak coherence of the light curves parallel/perpendicular to the shear.

<u>Numbers</u>. With a Hubble constant of $H_0 = 75$ km sec^{-1} Mpc^{-1} and a density parameter $q_0 = 0.5$, the Einstein radius in the source plane is

$$\xi_0 = (4GM/c^2)^{0.5}(D_{LS}D_S/D_L)^{0.5} = 4.19 \times 10^{-2} \times (m/M_\odot)^{0.5} \text{ pc}$$

(D_{LS}, D_S, and D_L are the angular diameter distances lens–source, observer–source, and observer–lens, respectively). The length of one pixel corresponds to 1 pixel = $\xi_0/12.5 = 3.35 \times 10^{-3}$ pc. For the time scale of the light curves, we used the values of KRS, $\Delta t = \xi_0/v_{eff}$, where the effective velocity v_{eff} corresponds to a center of mass velocity of the lensing galaxy of 600 km/sec. The crossing time for the region displayed in Figure 1a is of order 300 years (for details, see KRS). For other choices of the parameters, one can rescale the length and time scales.

<u>Discussion</u>. The light curves of the four images depend strongly on the relative directions of velocity and shear. Even for moderate source sizes, one obtains high peaks and deep valleys if the velocity is perpendicular to the action of the shear, whereas in the parallel case the light curves are much smoother. This influence is much stronger than the dependency of the different combinations of shear and surface density for the four images.

Changes of the luminosity of order 1 mag within about 5 years are quite common in our calculations for the source sizes smaller than about 0.015 pc. The differences in the observed

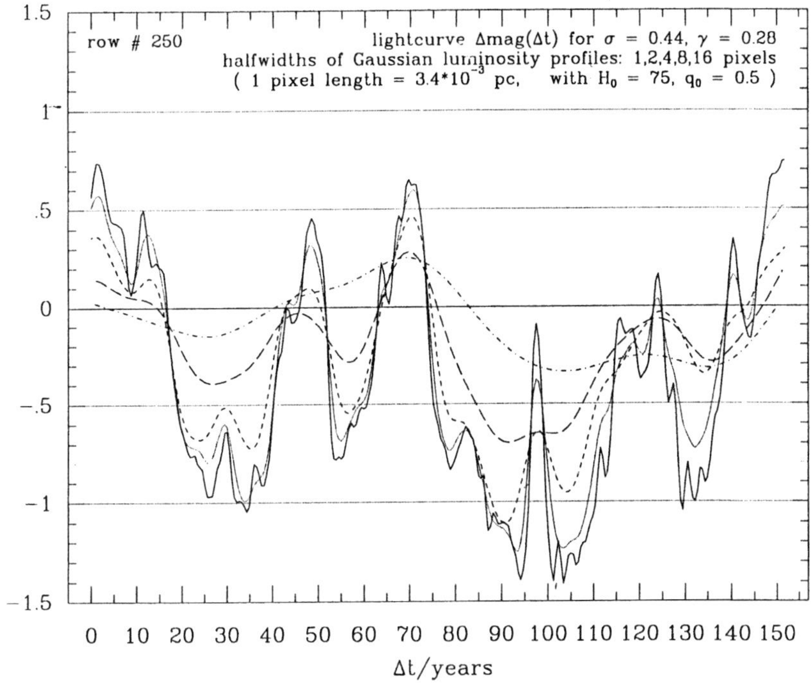

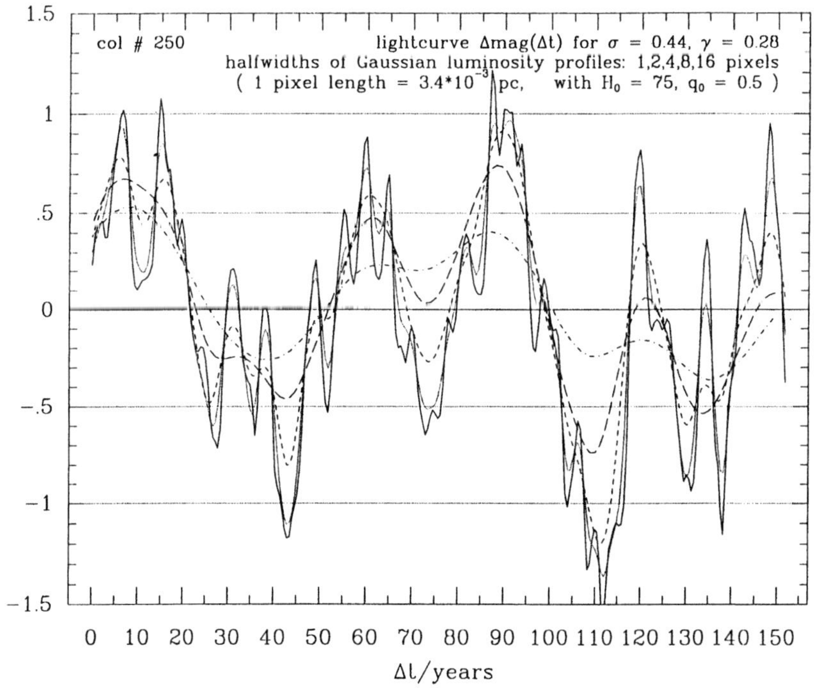

Figures 2a/b. Light curves L(t) for image B parallel ("row")/perpendicular ("col") to the action of the shear (path is indicated in Fig. 1a).

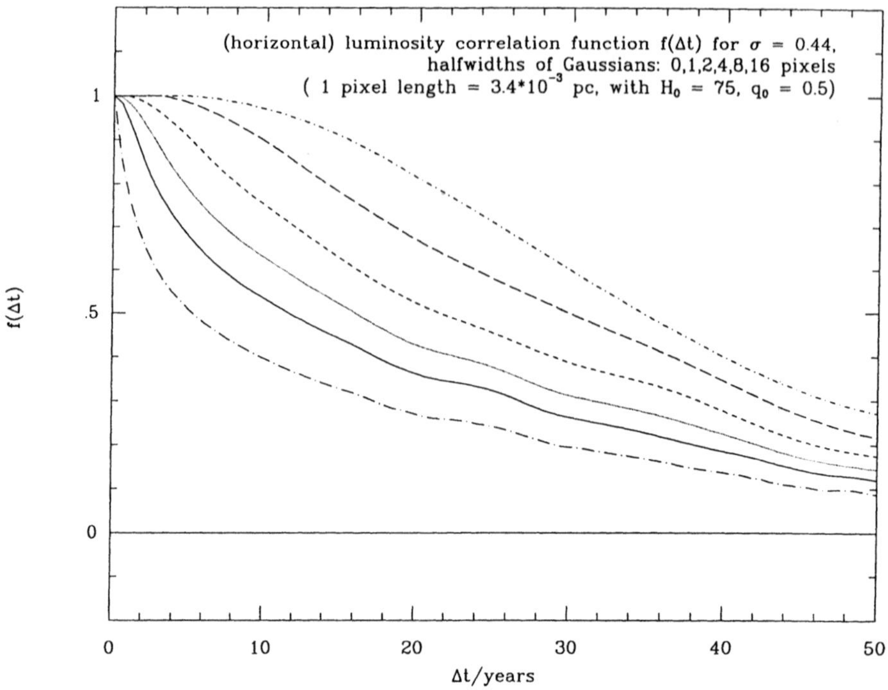

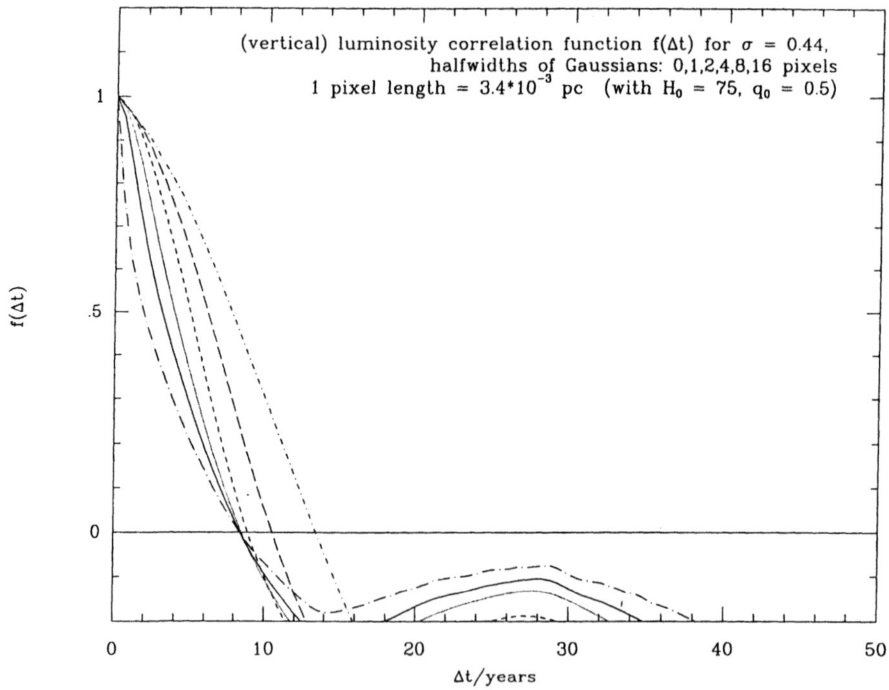

Figures 3a/b. Luminosity auto-correlation function f(Δt) for image B parallel ("horizontal")/ perpendicular ("vertical") to the action of the shear.

luminosities between the measurements in 1986 and 1987 (STGHSL) are −0.11, +0.24, −0.23, and −0.20 mag for the images A, B, C, D, respectively. Although the observed intensities may also be affected by interstellar extinction (cf. Yee 1988), we think is is not unreasonable to interpret these changes of luminosities as consequences of micro-lensing! A more detailed description of our results and of the computational technique will be given elsewhere.

This project was supported by NSF grant AST-85-19552 and NASA grant NAGW-765, and by a fellowship of the German Academic Exchange Service (DAAD) awarded to one of us (J.W.).

References.

Barnes, J., Hut, P. 1986, *Nature*, **324**, 446.

Katz, N., Balbus, S., Paczyński, B. 1986, *Ap. J.*, **306**, 2.

Kayser, R., Refsdal, S., Stabell, R. 1986, *Astr. Ap.*, **166**, 36.

Paczyński, B. 1986, *Ap. J.*, **301**, 503.

Schneider, D.P., Turner, E.L., Gunn, J.E., Hewitt, J.N., Schmidt, M., Lawrence, C.R. 1988, *A. J.*, **95**, 1619.

Schneider, P., Weiss, A. 1987, *Astr. Ap.*, **171**, 49.

Yee, H.K.C. 1988, *A. J.*, **95**, 1331.

Young, P. 1981, *Ap. J.*, **244**, 756.

A VIABLE EXPLANATION FOR QUASAR-GALAXY ASSOCIATIONS?

Peter Schneider
Max-Planck Institut für Astrophysik
Karl-Schwarzschild-Str. 1, D-8046 Garching, FRG

The Problem. For more than two decades, claims that there is an overdensity of quasars near foreground galaxies have been disputed in the literature (for a review, see Arp 1987). Undoubtedly, the associations found present large deviations from the average quasar density (e.g., Arp *et al.* 1984; Monk *et al.* 1988), but the statistical significance of these overdensities is difficult, if not impossible, to confirm. Whereas a small minority of astronomers take these observations as evidence for a noncosmological origin of (quasar) redshifts, they are mostly ignored by the majority of the astronomical community. Probably the main reason for the general disbelief of the statistical significance is the lack of an appealing explanation for the overdensities. Convinced of the cosmological origin for redshifts (for which there are many good arguments), one hardly can think of an effect that could yield a connection between high and low redshift objects.

Now, this is not really true, as we see absorption lines of vastly different redshift in the spectra of high redshift quasars. Not the object, but its photons can be influenced by intervening objects. Canizares (1981) proposed as a possible explanation for quasar-galaxy associations the gravitational deflection of the background quasar light by compact objects in the foreground galaxy. The possible magnification associated with the deflection can fool the observer: some quasars seen through this foreground galaxy are much fainter than he thinks they are. If he conducts a flux-limited quasar search, some of these sources are thus boosted in flux above the threshold of the sample. If this "amplification bias" is strong enough, one has a neat explanation for the "Arp effect," and, who knows, more effort and telescope time would be used to prove (or disprove) it. However, whereas several modifications of the original Canizares model have been published (e.g., Vietri and Ostriker 1983; Schneider 1986, 1987a,b; Peacock 1986), the bottom line remained basically unchanged (Canizares 1986): one has to search for quasars near a huge number ($> 10^4$) of galaxies to detect an overdensity. The reason for this is a conspiracy of two effects. For bright quasars, the source counts are relatively steep; hence, for every bright source, there is a large reservoir of faint sources that can be magnified, and thus a large fractional overdensity can occur. However, there are only few bright quasars in the sky, i.e., the statistics are bad. On the other hand, the statistics are much better for faint quasars, but there the source counts begin to flatten, the fractional overdensity is small. Since nobody has conducted a quasar search around that many galaxies, the overdensity (which undoubtedly is present in these lensing models) cannot have been detected at a level of statistical significance.

A Cure?. The implicit assumption made in the above cited lens models is that the overall source counts of quasars are not influenced by the amplification bias. That this is not necessarily the case has been pointed out by Barnothy and Barnothy (1968) and, in less radical terms, by Turner (1980), Canizares (1982), and others. A population of compact objects in the universe can significantly influence the source counts of high-redshift sources, if they are compact enough. In particular, if the high-L end of the luminosity function is sufficiently steep, the lens action can lead to source counts (at fixed redshift) that are considerably flatter. How flat? Well, if the source size is much smaller than the relevant length scales induced by the lenses (roughly $10^{16}\sqrt{M/M_\odot}$ cm) so that the sources can be effectively considered to be point sources, any lens-dominated source counts have to have an integral slope of -2 (Vietri 1985). But the above-quoted length scale is very comparable to the continuum emission size of quasars as estimated from the variability time scale of some of them. (This does of course not imply that all quasars are that compact.) If one takes into account the finite size of the sources, lens-dominated source counts can have a steeper slope (Schneider 1987c). In particular, it is possible to reproduce the observed slope (-2.6) if a suitable mass spectrum of the lenses is chosen.

The argument presented at the end of the first section then suggests the following idea (Linder and Schneider 1988): is it possible to "guess" an intrinsic quasar luminosity function and a lens population in such a way that, first, the overall source counts can be reproduced and, at the same time, a statistically significant overdensity near foreground galaxies can occur? The answer is partly provided in Figure 1 where for typical set of parameters, three curves are drawn: the "source counts" $\bar{n}$ that would be observed in a smooth Friedmann-Lemaitre universe (i.e., this curve reflects the intrinsic luminosity function of the sources, which is seen to be rather steep), the overall source counts n, and the source counts $\hat{n}$ within an arcmin of the center of a foreground galaxy. That is, the ratio between $\hat{n}$ and n is the fractional overdenity of quasars within an arcmin of the foreground galaxy. For the model of Figure 1, this overdensity is a factor of ≈ 30, and going through the statistical arguments, this is sufficient to provide an observable effect. That's the good news.

Note that the model parameters are rather realistic; for example, we have assumed that

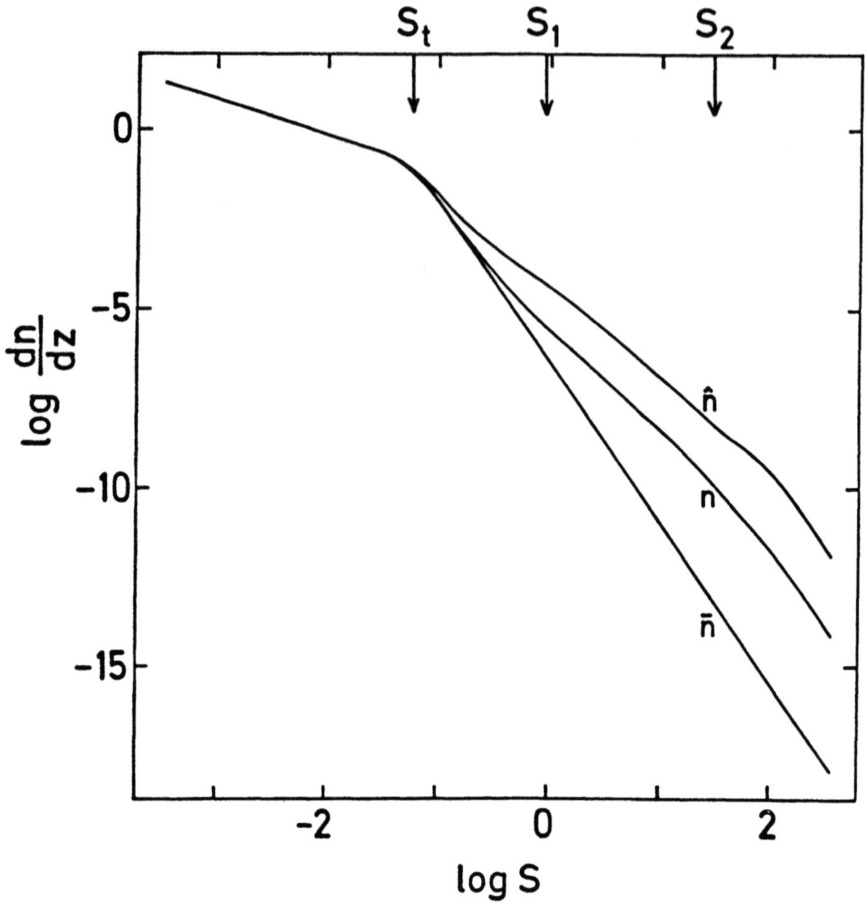

Fig. 1: Logarithm of source counts vs. logarithm of flux, both measured in arbitrary units. Counts in the absence of lensing (smooth universe) are denoted by $\bar{n}$, counts with lensing averaged over the sky by n, and those within one arcmin of a foreground galaxy by $\hat{n}$. A cosmological model with $\Omega = 1.0$ was assumed, and 5% of the mass is in compact objects; their mass distribution is a power law with the slope of the Salpeter IMF. The galaxy is assumed to have redshift $z_g = 0.01$ and velocity dispersion $\sigma_v = 300$ km/s. The sources are assumed to have redshift $z_s = 1.0$. S_t is the flux at which the unlenses source counts $\bar{n}$ turn over; between S_1 and S_2, the lensed source counts have an integral slope of -2.6.

only 5% of the critical mass in the universe is contained in the lens population of compact objects. Also, the high-S slope of the overall source counts n is nicely reproduced. The redshift and velocity dispersion of the foreground galaxy is about typical for the cases where overdensities are observed.

The bad news is that the overall source counts n have a feature that is not observed. Namely, if one extrapolates the high-S part of the n curve down to S_t where the source counts become flat, it falls below the actual value of $n(S_t)$ by about a factor of 20. Such a hump is not observed in the real world, and it is of such a magnitude that it would have been if it is there. The reason for the occurrence of the hump in our model is the following: the flat, low-S part of n is not influenced by lensing, whereas the high-S power law tail is determined by the lens population, its slope by the mass spectrum of the lenses, and its amplitude by their density. To get rid of the hump, one can increase the cosmological density of lenses, but then the fractional overdensity decreases. Or one can make the flat part of the luminosity function steeper, but that contradicts faint quasar counts and also violates the limits set by the X-ray background. Referring to the title of this paper, the answer is *no*.

In fact, one can raise further objections against the model presented. Clearly, its essential feature is that the overall source counts are lens dominated, i.e., basically all bright quasars are magnified. Referring to the length scales quoted above, this means that one would expect all these bright quasars to vary on a time scale on which the lensing parameters (relative alignment) changes, and this is typically on the order of a few years, and shorter for highly magnified sources (Gott 1981; Schneider and Weiss 1987). In reality, only a fraction of quasars is observed to vary substantially. Further arguments against lens-dominated source counts can be found in Canizares (1982).

A new piece of evidence *for* amplification bias is provided by the Surdej *et al.* (1988) sample (see the contribution by Turner in this volume): the fact that the fraction of multiply imaged quasars in a sample of high-luminosity sources is as high as they observe, much higher than the canonical fraction $(\text{few hundred})^{-1}$ (Zwicky 1937; Turner *et al.* 1984) shows that the amplification bias *is* operating at least for the brightest quasars in the sky, quite in agreement with the model by Ostriker and Vietri (1986). Otherwise, one could not understand why there is a correlation between the flux of a source and the probability of its being multiply imaged. On the other hand, lens searches at somewhat fainter magnitude (Webster *et al.* 1988) turn out to be less "successful," i.e., the fraction of multiply imaged sources is much smaller than in the Surdej *et al.* (1988) sample. Hence, the amplification bias *only* operates for the brightest quasar magnitudes, but most probably not for the average 19th magnitude, $z \sim 2$ quasar. This of course is in marked contrast to the model of Figure 1 where one has an appreciable range in flux over which the source counts are lens dominated.

Discussion. The arguments outlined above show that the Arp effect can not be explained by gravitational lensing by compact objects in the halos of foreground galaxies, even if one drops the assumption that the overall source counts are nearly unaffected by the amplification bias. Needless to say, this conclusion will trigger a reaction: "Well, there was nothing to explain in the first place." Thus, the original microlensing problem, although well understood, seems to lack an application.

On the other hand, progress is being made in observations of quasar samples to look for multiple images. Once statistically well defined samples are observed and analyzed, we will obtain important information on the lens population in the universe. Just to give an example: suppose the fraction of multiply imaged quasars in a complete sample of high-luminosity quasars is ν, much larger than the fraction expected if the amplification bias were absent. If the emission region of quasars is indeed smaller than the relevant length scale induced by stellar mass lenses, then the amplification bias is due to microlensing. In order to couple microlensing (amplification) to macrolensing (image splitting), the microlenses must be associated with the macrolenses. In other words, one can conclude that about a fraction ν of all microlenses lies within the critical radius of macrolenses for producing macroimages. Now, this argument only works if the emitting region of quasars is indeed small. Otherwise, the amplification bias is produced by the macrolenses themselves. In this case, the radio samples should show a similar behavior, namely, that the most luminous radio quasars should have a relatively large probability of being multiply imaged. A large quasar size would also help to understand why we haven't seen a clear indication of microlensing through variability in one of the known multiple imaged sources. For example, Young (1981) concluded from the lack of rapid variability in 0957+561 that the emission region has a size in excess of 10^{16} cm.

The fact that the overdensity of quasars near galaxies cannot be explained by lensing does not mean that an observed overdensity of galaxies around high redshift quasars (Webster, this volume) is also not attributable to microlensing. The reason for the intrinsic difference of these two questions is that quasars are rare and galaxies are not. The bad statistics, which kill the lens-induced Arp effect, will have a much smaller impact on the reverse effect. Surprisingly, this question has not yet been addressed theoretically.

It is a pleasure to thank J. Gunn and J. Ostriker for helpful discussions.

References.

Arp, H. 1987, *Quasars, Redshifts, and Controversies* (Interstellar Media, Berkeley).

Arp, H., Wolstencroft, R.D., and He, X.T. 1984, *Ap. J.*, **285**, 44.

Barnothy, J., and Barnothy, M.F. 1968, *Science*, **162**, 348.

Canizares, C.R. 1981, *Nature*, **291**, 620.

Canizares, C.R. 1982, *Ap. J.*, **263**, 508.

Canizares, C.R. 1986, in *Observational Cosmology, Proceedings of IAU Symposium No. 124*, ed. A. Hewitt, G. Burbidge, and L.Z. Fang (Dordrecht: Reidel), p. 729.

Gott, J.R. 1981, *Ap. J.*, **243**, 140.

Linder, E.L., and Schneider, P. 1988, *Astr. Ap.*, (in press).

Monk, A.S. *et al.* 1988 (preprint).

Ostriker, J.P., and Vietri, M. 1986, *Ap. J.*, **300**, 68.

Peacock, J.A. 1986, *M. N. R. A. S.*, **223**, 113.

Schneider, P. 1986, *Ap. J. (Letters)*, **300**, L7.

Schneider, P. 1987a, *Astr. Ap.*, **179**, 71.

Schneider, P. 1987b, *Astr. Ap.*, **179**, 80.

Schneider, P. 1987c, *Astr. Ap.*, **183**, 189.

Schneider, P., and Weiss, A. 1987, *Astr. Ap.*, **171**, 49.

Surdej, J., Swings, J. P., Magain, P., Borgeest, U., Kayser, R., Refsdal, S., Courvoisier, T. J. L., Kellermann, K. I., and Kuhr, H. 1988, in *Proceedings of a Workshop on Optical Surveys for Quasars*, ed. P. S. Osmer, A. C. Porter, R. F. Green, and C. B. Foltz (Astronomical Society of the Pacific, San Francisco), p. 183.

Turner, E.L. 1980, *Ap. J. (Letters)*, **242**, L135.

Turner, E.L., Ostriker, J.P., and Gott, J.R. 1984, *Ap. J.*, **284**, 1.

Vietri, M. 1985, *Ap. J.*, **293**, 343.

Vietri, M., and Ostriker, J.P. 1983, *Ap. J.*, **267**, 488.

Webster, R.L., Hewett, P.C., and Irwin, M.J. 1988, *A. J.*, **95**, 19.

Young, P. 1981, *Ap. J.*, **244**, 756.

Zwicky, R. 1937, *Phys. Rev.*, **51**, 679.

Reception Photographs

Claire Burke Jane Burke Mark Burke Bernard Burke

Marc Gorenstein Bernard Burke Albert Hill

Dennis Walsh Bernard Burke Cosmo Papa

Fred Lo Mark Reid Paul Vanden Bout

Marc Gorenstein John Tonry Charles Lawrence Martin Ewing

Woody Strandberg Hans Hinteregger

Ted Reifenstein Philip Rosenkrantz Martin Ewing

Albert Hill David Frisch

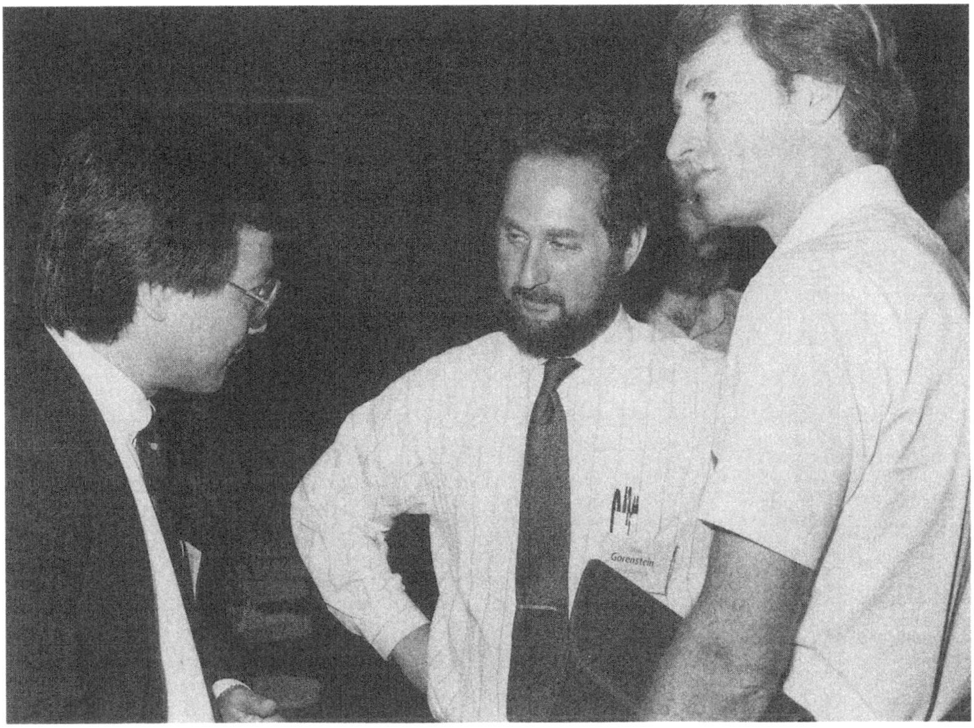

Fred Lo   Marc Gorenstein   Charles Lawrence

James Wright   Joseph Salah

James Wright Jacqueline Hewitt

Jean Andersen Bernard Burke

Program
for
Conference on Gravitational Lenses

June 20, 1988
Bush Room (10-105)
M.I.T.
77 Massachusetts Avenue
Cambridge, MA 02139

8:15 Registration/Poster Setup

9:00 Welcome
George Clark
Announcements
James Moran

Chairperson (morning session) *K. Y. Lo*

9:15 0957+561: The Unpublished Story
Dennis Walsh, NRAL, Jodrell Bank

9:45 The Versatile Elliptical Gravitational Lens
Ramesh Narayan and Scott Grossman, University of Arizona

10:00 Gravitational Lens Optics
Roger Blandford, CalTech

10:30 Posters/Refreshments

10:45 VLBI Observations of Gravitational Lenses
Marc Gorenstein, CfA

11:15 VLBI Phase Reference Mapping Techniques and the Search for the Third Image of 0957+561
Alan Rogers, Haystack Observatory

11:30 An Estimate of the Time Delay in the Gravitationally Lensed Double Quasar 0957+561
Joseph Lehár, MIT

11:45 The VLA Gravitational Lens Survey: An Update
Jacqueline Hewitt, Haystack Observatory

12:15 Group Photo, Lunch

Chairperson (afternoon session) *Jacqueline Hewitt*

13:45 Recent Optical Observations of Gravitational Lenses
Edwin Turner, Princeton University Observatory

14:15 Optical Searches for Gravitational Lenses
Rachel Webster, University of Toronto

14:45 Gravitational Lens Model for Arcs in Clusters of Galaxies
Vahé Petrosian, Stanford University
15:15 A Gravitational Telescope in A370: Indeed It Works!
Genevieve Soucail, Toulouse Observatory

15:30 Posters/Refreshments

15:45 Gravitational Microlensing
William Watson, University of Illinois
16:15 A Viable Explanation of Quasar-Galaxy Associations
Peter Schneider, MPI Astrophysics
16:30 Statistics of Gravitational Lenses: Galaxies and Dark Matter
Lawrence Krauss, Yale University
16:50 Highly Colinear Radio Sources and Constraints on Gravitational Lens Space Density
Colin Lonsdale, Haystack Observatory
17:00 Explaining Burke and Shapiro to Newton
David Frisch, MIT

17:15 Adjourn

18:30 Banquet, Hyatt Regency Hotel

Poster Session

1 History of Quasars as Gravitationally Lensed Images
Jeno Barnothy
2 Investigating Large-Scale Structure Using a Thin Lens Approximation in a Nonhomogeneous Cosmological Model
Daniel Blanchard and Charles Dyer, University of Toronto
3 Observations of the Time Delay in the Double Quasar 0957+561
J. Schneider and Genevieve Soucail, Toulouse Observatory
4 Is the Giant Luminous Arc Due to Lensing by a Cosmic String?
Wu Xiangping, Beijing Astronomical Observatory
5 Resolution of the Gravitational Lens of Quasar 2016+112: Measurement of a Third Image Flux and Position
Glen Langston, NRL, C. Carrilli, Samuel Conner, Michael Heflin, Bernard Burke, MIT, Jacqueline Hewitt, Haystack, and Charles Lawrence, CalTech
6 Gravitational Lensing of Extended Sources
S. M. Chitre, Institute of Astronomy, and D. Narasimha, TIFR, Bombay
7 Observations of the Blue Arcs in Abell 963
Russell Lavery, University of Hawaii
8 Quasar Microlensing and Dark Matter
Hans-Walter Rix and Craig Hogan, University of Arizona
9 Microlensing Model for 2237+0305
Joachim Wambsganss, Bohdan Paczynski, and N. Katz, Princeton University Observatory
10 Moving Gravitational Lenses
Mark Birkinshaw, CfA

Author Index

Source Index

Page numbers refer to first page of article.

Subject Index

Page numbers refer to first page of article.

Lecture Notes in Mathematics

Vol. 1236: Stochastic Partial Differential Equations and Applications. Proceedings, 1985. Edited by G. Da Prato and L. Tubaro. V, 257 pages. 1987.

Vol. 1237: Rational Approximation and its Applications in Mathematics and Physics. Proceedings, 1985. Edited by J. Gilewicz, M. Pindor and W. Siemaszko. XII, 350 pages. 1987.

Vol. 1250: Stochastic Processes – Mathematics and Physics II. Proceedings 1985. Edited by S. Albeverio, Ph. Blanchard and L. Streit. VI, 359 pages. 1987.

Vol. 1251: Differential Geometric Methods in Mathematical Physics. Proceedings, 1985. Edited by P.L. García and A. Pérez-Rendón. VII, 300 pages. 1987.

Vol. 1255: Differential Geometry and Differential Equations. Proceedings, 1985. Edited by C. Gu, M. Berger and R.L. Bryant. XII, 243 pages. 1987.

Vol. 1256: Pseudo-Differential Operators. Proceedings, 1986. Edited by H.O. Cordes, B. Gramsch and H. Widom. X, 479 pages. 1987.

Vol. 1258: J. Weidmann, Spectral Theory of Ordinary Differential Operators. VI, 303 pages. 1987.

Vol. 1260: N.H. Pavel, Nonlinear Evolution Operators and Semigroups. VI, 285 pages. 1987.

Vol. 1263: V.L. Hansen (Ed.), Differential Geometry. Proceedings, 1985. XI, 288 pages. 1987.

Vol. 1265: W. Van Assche, Asymptotics for Orthogonal Polynomials. VI, 201 pages. 1987.

Vol. 1267: J. Lindenstrauss, V.D. Milman (Eds.), Geometrical Aspects of Functional Analysis. Seminar. VII, 212 pages. 1987.

Vol. 1269: M. Shiota, Nash Manifolds. VI, 223 pages. 1987.

Vol. 1270: C. Carasso, P.-A. Raviart, D. Serre (Eds.), Nonlinear Hyperbolic Problems. Proceedings, 1986. XV, 341 pages. 1987.

Vol. 1272: M.S. Livšic, L.L. Waksman, Commuting Nonselfadjoint Operators in Hilbert Space. III, 115 pages. 1987.

Vol. 1273: G.-M. Greuel, G. Trautmann (Eds.), Singularities, Representation of Algebras, and Vector Bundles. Proceedings, 1985. XIV, 383 pages. 1987.

Vol. 1275: C.A. Berenstein (Ed.), Complex Analysis I. Proceedings, 1985–86. XV, 331 pages. 1987.

Vol. 1276: C.A. Berenstein (Ed.), Complex Analysis II. Proceedings, 1985–86. IX, 320 pages. 1987.

Vol. 1277: C.A. Berenstein (Ed.), Complex Analysis III. Proceedings, 1985–86. X, 350 pages. 1987.

Vol. 1283: S. Mardešić, J. Segal (Eds.), Geometric Topology and Shape Theory. Proceedings, 1986. V, 261 pages. 1987.

Vol. 1285: I.W. Knowles, Y. Saitō (Eds.), Differential Equations and Mathematical Physics. Proceedings, 1986. XVI, 499 pages. 1987.

Vol. 1287: E.B. Saff (Ed.), Approximation Theory, Tampa. Proceedings, 1985–1986. V, 228 pages. 1987.

Vol. 1288: Yu. L. Rodin, Generalized Analytic Functions on Riemann Surfaces. V, 128 pages, 1987.

Vol. 1294: M. Queffélec, Substitution Dynamical Systems – Spectral Analysis. XIII, 240 pages. 1987.

Vol. 1299: S. Watanabe, Yu.V. Prokhorov (Eds.), Probability Theory and Mathematical Statistics. Proceedings, 1986. VIII, 589 pages. 1988.

Vol. 1300: G.B. Seligman, Constructions of Lie Algebras and their Modules. VI, 190 pages. 1988.

Vol. 1302: M. Cwikel, J. Peetre, Y. Sagher, H. Wallin (Eds.), Function Spaces and Applications. Proceedings, 1986. VI, 445 pages. 1988.

Vol. 1303: L. Accardi, W. von Waldenfels (Eds.), Quantum Probability and Applications III. Proceedings, 1987. VI, 373 pages. 1988.

Lecture Notes in Physics

Vol. 305: K. Nomoto (Ed.), Atmospheric Diagnostics of Stellar Evolution: Chemical Peculiarity, Mass Loss, and Explosion. Proceedings, 1987. XIV, 468 pages. 1988.

Vol. 306: L. Blitz, F.J. Lockman (Eds.), The Outer Galaxy. Proceedings, 1987. IX, 291 pages. 1988.

Vol. 307: H.R. Miller, P.J. Wiita (Eds.), Active Galactic Nuclei. Proceedings, 1987, XI, 438 pages. 1988.

Vol. 308: H. Bacry, Localizability and Space in Quantum Physics. VII, 81 pages. 1988.

Vol. 309: P.E. Wagner, G. Vali (Eds.), Atmospheric Aerosols and Nucleation. Proceedings, 1988. XVIII, 729 pages. 1988.

Vol. 310: W.C. Seitter, H.W. Duerbeck, M. Tacke (Eds.), Large-Scale Structures in the Universe – Observational and Analytical Methods. Proceedings, 1987. II, 335 pages. 1988.

Vol. 311: P.J.M. Bongaarts, R. Martini (Eds.), Complex Differential Geometry and Supermanifolds in Strings and Fields. Proceedings, 1987. V, 252 pages. 1988.

Vol. 312: J.S. Feldman, Th.R. Hurd, L. Rosen, "QED: A Proof of Renormalizability." VII, 176 pages. 1988.

Vol. 313: H.-D. Doebner, T.D. Palev, J.D. Hennig (Eds.), Group Theoretical Methods in Physics. Proceedings, 1987. XI, 599 pages. 1988.

Vol. 314: L. Peliti, A. Vulpiani (Eds.), Measures of Complexity. Proceedings, 1987. VII, 150 pages. 1988.

Vol. 315: R.L. Dickman, R.L. Snell, J.S. Young (Eds.), Molecular Clouds in the Milky Way and External Galaxies. Proceedings, 1987. XVI, 475 pages. 1988.

Vol. 316: W. Kundt (Ed.), Supernova Shells and Their Birth Events. Proceedings, 1988. VIII, 253 pages. 1988.

Vol. 317: C. Signorini, S. Skorka, P. Spolaore, A. Vitturi (Eds.), Heavy Ion Interactions Around the Coulomb Barrier. Proceedings, 1988. X, 329 pages. 1988.

Vol. 319: L. Garrido (Ed.), Far from Equilibrium Phase Transitions. Proceedings, 1988. VIII, 340 pages. 1988.

Vol. 320: D. Coles (Ed.), Perspectives in Fluid Mechanics. Proceedings, 1985. VII, 207 pages. 1988.

Vol. 321: J. Pitowsky, Quantum Probability – Quantum Logic. IX, 209 pages. 1989.

Vol. 322: M. Schlichenmaier, An Introduction to Riemann Surfaces, Algebraic Curves and Moduli Spaces. XIII, 148 pages. 1989.

Vol. 323: D.L. Dwoyer, M.Y. Hussaini, R.G. Voigt (Eds.), 11th International Conference on Numerical Methods in Fluid Dynamics. XIII, 622 pages. 1989.

Vol. 324: P. Exner, P. Šeba (Eds.), Applications of Self-Adjoint Extensions in Quantum Physics. Proceedings, 1987. VIII, 273 pages. 1989.

Vol. 327: K. Meisenheimer, H.-J. Röser (Eds.), Hot Spots in Extragalactic Radio Source. Proceedings, 1988. XII, 301 pages. 1989.

Vol. 328: G. Wegner (Ed.), White Dwarfs. Proceedings, 1988. XIV, 524 pages. 1989.

Vol. 329: A. Heck, F. Murtagh (Eds.), Knowledge Based Systems in Astronomy. IV, 280 pages. 1989.

Vol. 330: J.M. Moran, J.N. Hewitt, K.Y. Lo (Eds.), Gravitational Lenses. Proceedings, 1988. XIV, 238 pages. 1989.

GPSR Compliance

The European Union's (EU) General Product Safety Regulation (GPSR) is a set of rules that requires consumer products to be safe and our obligations to ensure this.

If you have any concerns about our products, you can contact us on ProductSafety@springernature.com

In case Publisher is established outside the EU, the EU authorized representative is:

Springer Nature Customer Service Center GmbH
Europaplatz 3
69115 Heidelberg, Germany

Batch number: 10620881

Printed by Printforce, the Netherlands